History of Regional Science
and the
Regional Science Association International

Springer

Berlin
Heidelberg
New York
Hong Kong
London
Milan
Paris
Tokyo

Walter Isard

History of Regional Science and the Regional Science Association International

The Beginnings and Early History

Springer

Professor Dr. Walter Isard
Cornell University
Department of Economics
Uris Building
Ithaca, New York 14583
USA

ISBN 3-540-00934-5 Springer-Verlag Berlin Heidelberg New York

Cataloging-in-Publication Data applied for

A catalog record for this book is available from the Library of Congress.

Bibliographic information published by Die Deutsche Bibliothek
Die Deutsche Bibliothek lists this publication in the Deutsche Nationalbibliografie; detailed bibliographic data available in the internet at *http.//dnb.ddb.de*

Springer-Verlag Berlin Heidelberg New York
a member of BertelsmannSpringer Science + Business Media GmbH

http://www.springer.de

Printed in Germany

Cover design: Erich Kirchner, Heidelberg

SPIN 10922644 42/3130 – 5 4 3 2 1 0 – Printed on acid-free paper

Acknowledgments

Once again I am grateful to both Helena Wood and Henning Pape-Santos for excellent work on my almost illegible manuscript. In the writing of this book I have received continued support and encouragement from David E. Boyce, Kieran P. Donaghy and Geoffrey J. D. Hewings. David Boyce, as Archivist of the Regional Science Association International, has made a number of suggestions that have led to significant improvement in the organization of the book and its coverage of materials. He also succeeded me in carrying forward the effective management and organization of the Association when such advancement was beyond my capabilities. As with so many times before, my wife, Caroline, has graciously put up with my extreme writing sprees.

Preface

It is difficult, if not impossible, to establish the point of time at which a new field of study starts to emerge. While the date of formal organization of a society associated with the field can be precisely stated, such timing says little about when and where the seeds for a field's development were planted. Also, such timing says little about the essential "why" for the development of a field and provides little understanding of the path that it traced.

It is clear that the emergence of the field of regional science, like many other fields, was dependent on a particular setting as well as the pattern of events and interaction of diverse personalities who became involved.

As best I can, I shall attempt in this Part I of the History to unravel the *where, when* and *why* questions in the development of Regional Science and the Regional Science Association International. Further, in the last section of this essay, I shall briefly point up some potentialities for analytical advances in the field of regional science. Also, I shall note the opportunity for leadership by regional scientists in attacking global and regional development problems, thereby to help formulate relevant policy. In this way, I hope to expose the potential for fruitful research by young scholars interested in entering the field.

Walter Isard
April 2003

Table of Contents

1 The Setting and Initial Events

There are at least three basic factors that must be emphasized in accounting for the development of regional science and the formation of the Regional Science Association International, RSAI. One is the state of affairs at the start of the period when regional science began to develop. Another concerns the events, processes and structural elements of the world that conditioned and led to the entrepreneurial activity of relevant personalities. Still another must account for the existence and provision of resources and the potential for a scientific thrust that made possible the development of a new field of study.

Let us begin with the state of affairs immediately after World War II that characterized the social science areas and their interest in and treatment of regional analysis and problems. At that time and in the immediate years that followed there was a sudden increase in the intensity of interest in regional analysis. In the United States, as in many other parts of the world, many regional problems had been either generated or aggravated by the war. Disruptions and displacements were common. At that time, city planning as a profession, which before the war had been seriously questioned by social scientists and laymen in North America, began to be widely accepted. Still, however, no forum had evolved at which urban and regional problems, in their several dimensions, could be discussed. The various traditional disciplines did not become seriously concerned with these problems until the last half of the 1960s. In the field of economics, location theory was something perceived to be as far away from the center of economics as the farthest planet is from the sun. There were no urban and regional economics. Since economists conducting regional studies were looked down upon, less than a handful of good ones were in the field; similarly, local and urban affairs were belittled areas for specialization in political science. Regional sociology, it is true, had gained wide acceptance and considerable prestige at the hands of Howard Odum, Rupert Vance, and their associates at the University of North Carolina. But, as no young scholar had emerged to provide the dynamic leadership that had been exercised by Odum, this new area of study had begun to decline. Geography was not at a high level, academically speaking, being primarily concerned at that time with cartography, the gathering of data and their simple processing, with little concern for analysis, except of the simplest type. I well recall the standard articles on iron and steel location. They dealt with factors such as availability and quality of ore, coal, limestone, labor, and other inputs, the accessibility to different markets, and the quality of management. No attempt was made on the part of the economic geographers to fuse all these factors into a simple cost calculus, as could easily be done. I remember the resistance to my attempt to do so in a paper presented at the meetings of the Association of American Geographers in Worcester in 1950.[1]

So, to repeat, an urgent need for urban and regional analysis had developed, with practically no social science discipline willing to nurture it.

Into the above situation an entrepreneurial personality with particular background and training emerged. The direction that that personality took as well as the provision of resources that were required in the development of the field of regional science were greatly conditioned by four changes that had occurred in the world situation. These were:

- The Decline of the British Empire
- The Rise of Hitler
- The Introduction and later Widespread Use of Mathematics in Economics
- The Keynesian Revolution

Let us start with the relevance of the first.

In 1919 when a son of immigrant parents was born, to these parents the peak of intellectual activity in the world was at Oxford University. They aspired for that son to attend that university or other leading one in Britain; they had not yet known or perceived that the mighty British Empire was already falling. (According to the historians, Roberts and Roberts (1985), this empire controlled in 1900 almost 25 percent of the land of the Globe and 25 percent of its population, p. 642.) By 1939 the decline was so advanced that Oxford had been clearly replaced by Harvard as the leading university in the field of economics. In that year that son of immigrant parents had been urged by his economics professor to attend Harvard for graduate study. Why was this important? One should recall the extreme bias of Anglo-Saxon economists existing then and previously. Location and spatial analysis had no place in their perception of the world and how it functioned. Most of them (if not all) were living in a world that geographers would say was a wonderland of no dimensions. It must be admitted that Marshall had on occasion casually to mention place and location in his Principles of Economics, but such had no effect on his analysis and that of the distinguished set of economists, Edgeworth, Wicksteed, Pigou and others, who were his contempories and who followed. Had that son of immigrant parents attended Oxford or a sister British university, study of location and space in all likelihood would have been squashed. As late as 1966,

[1] When I submitted the paper for publication in *Economic Geography,* the leading journal in that field, I had to fight vigorously to persuade the editors to permit me to retain the footnotes and citations essential to scholarship.

every paper that he had submitted for publication in British journals had been rejected.

The second of the four interconnected changes was the rise of Hitler. The 1929 Stock Market Crash and the following depression had impacted the German economy most disastrously. This was in large part the result of the reparation payments exacted from Germany by the Allies after World War I. The ensuing election of Adolf Hitler as Chancellor, and the rise of the Nazi party to power coupled with Hitler's notion of Aryan superiority and associated geopolitical doctrine was the death-knell of all effective intellectual activity in German economics. The advance of the Historical School that had developed in Germany and which had nourished location analysis was brought to a halt. The balance that its literature had provided to Anglo-Saxon non-spatial writings was destroyed. Thus, the field for significant advance in locational thought became wide-open. It is true that August Lösch had published his *Die räumliche Ordnung der Wirtschaft* in 1940, but along with his penniless condition that work was ignored; it was hardly known in the English speaking countries. Likewise, Christaller's 1935 central place study was given short shrift.

The third major factor was the introduction of mathematics into economics and its subsequent widespread employment in this field. Its significance for regional science will be presented later.

The fourth major change was the Keynesian Revolution, brought about by the 1936 publication of Manyard Keynes' book, *General Theory of Employment, Interest and Money.* It led to the extensive development of macroeconomics and the believed need for government deficit spending to combat depression. The principal proponent of that theory in the United States was Alvin H. Hansen who was appointed to the Lucious N. Littauer Professorship in Political Economy at Harvard University and its newly established Littauer Center. Also, Hansen was the chief economic advisor to President Franklin Delano Roosevelt, and during the school year he would spend two days a week in Washington, D.C.

In 1939 the unemployment rate in United States was still at a very high level, namely 17.2 percent. Hansen, being influenced by the pre-war structuralist doctrine of the German economist Spiethoff, attributed the sorry state of the U.S. economy to stagnation of investment—in industrial plant and equipment, housing and other construction, transport development and like activity. In 1940-41 the son of immigrant parents, in his second year of graduate study at Harvard, enrolled in the celebrated Hansen/Williams Fiscal Policy seminar, which most students, including the best, attended. Professor John Williams, who also served as the Dean of the Littauer School and Vice-President for Research at the Federal Reserve Bank in New York, provided a cautious, conservative approach to monetary and fiscal policy. That approach balanced the vigorous, more revolutionary doctrine of

Hansen—in a way that was polite yet made the seminar tremendously stimulating to a graduate student—truly a once-in-a-lifetime experience.

All enrollees of the seminar were expected to present a paper on their research. This son chose to do so on the cycle in apartment construction, later expanded to cover all building activity. In the previous year he had attended the Economic History course of Professor Abbott P. Usher. There Usher attributed the rise in the early Renaissance period of the Italian port cities largely to the fertile land of their hinterlands. Tillers of this land were able to provide more food than that required by their families. Thus, they could very easily produce a surplus to support the families of others choosing to engage in crafts, government, finance, religious activities and other occupations required for the growth of a trading city. Other excellent port sites with no access to fertile hinterland, were not able to develop into prominence because of high transport cost on food that would be required by an urban population.

In writing on the building cycle for the Fiscal Policy seminar, the son immediately recognized an obvious causal-type relationship between transport development and building.[2] Transport developments, those in the United States, namely, (1) the canal in the late 1820s up to the 1840s, (2) the early local railroads from the very late 1840s through the 1850s, (3) the interregional transport development from the late 1860s to the early 1870s, (4) the transcontinental rail development from the late 1870s through the 1880s, and finally (5) the street and electric railway construction boom in the late 1890s and first decade of the 20th century, each led to the opening up of new areas for economic development. Each lowered costs of raw materials and other inputs for production by firms and costs of consumption for population at developed centers. More important, each greatly stimulated trade and production at the new terminals located in the opened-up areas and at strategic sites along the new transport routes. In turn, each led to service trade development and major demand for housing and service activities at all the interconnected intermediate and terminal sites.[3]

---

[2] When I prepared my paper, I did have an opportunity to read the work on building cycles of Jan Tinbergen (1939) (who later shared the first Nobel Prize in Economics with Ragnar Frisch). In true Anglo-Saxon fashion he emphasized monetary factors. But it was clear that, despite the importance of the interest rate and other monetary factors, Tinbergen failed (at least for analyzing the lapse of building in the United States) to identify basic structural factors which were so much more important in Hansen's thinking.

[3] As examples: (1) the Erie canal led to the tremendous growth of Buffalo, and a number of cities along its course; (2) each major spurt in railroad construction led to a spurt in the development of Chicago, a major terminal as well as a major transshipment point; (3) the street and electric railway boom opened up outlying areas of major cities for families wishing to locate outside the more congested central and other districts. Thus there resulted housing booms in these new trade centers, transshipment points, intermediate cities and in outlying areas and significant investment in construction and diverse services.

To Hansen, the location analysis linking in large part the lapse in building activity in the 1930's to the lack of major transport development added considerable strength to the structural basis of his stagnation thesis, and hence for his associated policy recommendation for more government deficit spending. Thus, despite the Anglo-Saxon bias toward location and spatial analysis that carried over to and characterized the economics faculty at Harvard, Chicago and other leading universities,[4] the importance of location analysis, via Hansen's forceful presentations and his prominence in national policy formulation, became recognized in fiscal policy thinking.

Consequent to that son's pointing up the above obvious relation, Hansen strongly recommended for publication that son's research papers and, equally important, supported his applications for fellowships. In year 1942-43, two years after he matriculated at Harvard, he was awarded his first graduate fellowship, one to study at the University of Chicago. (Harvard, dominated by conservative economic professors, never supplemented with financial aid his earnings from piano playing at night spots, some of which definitely did not cater to the carrier trade.) With a generous fellowship award and the lure of studying with the famous Frank H. Knight, a strong anti–deficit spending advocate, that son sampled and absorbed the doctrines of that second most distinguished economics department. There he took a course on mathematical economics with Oskar Lange, a superb teacher and a leading light of the Cowles Commission in Economics. This was important for the later development of regional science. To explain why, it is to be noted that that son, as an undergraduate at Temple University, a poor man's college, had chosen to major in mathematics. However, the teaching of that subject was so inadequate, that in frustration he switched his major to economics, and thereafter matriculated for graduate study in economics at Harvard. Lange's, point by point, lucid presentation of the new mathematical economics, made it so easy to follow and so clear that he rekindled that son's interest in mathematics. In later years that son took over the Lange form of presentation that enabled him to require of regional science students understanding and use of the same advanced mathematical models and theories. (For example, the famous and highly advanced Arrow and Debreu (1954) existence theorem became required reading.) This mathematics

---

[4] For example, when I spoke to Joseph A. Schumpeter regarding transport cost as an important factor in location analysis, his highly unimaginative reply was that transport cost is just another production cost and that its impact can be covered by standard production cost analysis. In developing his classic and basic economics textbook, first edition 1955, Paul A. Samuelson did consider that possibly he should say something about the location problem in this book, and did query me about the subject, but in the end made no mention of it. Hicks, Fellner, Habeler, Machlup, and other leading economists (excluding Edgar M. Hoover, and Wolfgang F. Stolper) similarly did not consider location problems or analysis worthy of serious attention.

requirement, along with the superbly organized and meticulous editing of Ronald Miller, was largely responsible for the Journal of Regional Science, the outlet for papers of the regional science faculty and their former students, to receive a very high ranking among economic journals and others in closely related fields— exceedingly high for a journal that had been in existence for less than a dozen years.

To return to the Chicago days, this son also attended Jacob Viner's first year graduate course on economic theory. There Viner vigorously presented outstanding lectures on production costs, employing cost curves in a highly skilled manner for illuminating the interplay of fixed and variable costs for both the short-run and long-run. This proved extremely important for the research to be mentioned later.

2 The Emergence and Struggling Years of Regional Science

2.1 Early Developments and Significant Recognition of Location Theory

The following academic year, 1943–44, this son spent in Washington, D.C. Based upon a strong recommendation by Professor Alvin H. Hansen he had been awarded a highly coveted Social Science Research Council fellowship. To pursue further his location analysis and studies, he was attached to the National Resources Planning Board in Washington D.C. His advisor was Ernest Williams with whom he had stimulating discussions; and as a result he came to view Location and Transportation as the two sides of the same coin. Transportation analysis deepens location analysis and vice versa, and as a consequence both transportation and location models became basic in later regional science research.

While at the National Resources Planning Board he had occasions to discuss with Robert B. Mitchell and G. Holmes Perkins (who later spawned the Golden Age of City Planning at the University of Pennsylvania) the implications of both location and transportation analysis for urban development and planning. In these discussions it became obvious that there was a major overlap of interest. Location and transportation analysis can enhance city and regional planning and that planning can be, and later on did become, extremely valuable in pointing up critical and fruitful areas for location and transportation research, especially when conducted jointly with planning analysts.

In 1944, as a conscientious objector to military activity, this son was drafted and assigned to Civilian Public Service. One assignment was to serve as an orderly on the night shift at the State Hospital (a mental institution) in Middletown, Connecticut. There in the wee hours of the morning, when the many patients happened to be asleep and all was quiet on the wards, he took the occasion to translate into English the German literature on location—the works of Lösch, Weigmann, Engländer, Predöhl and others—and was able to grasp fully the scope and depth of this literature. For the English speaking community a summary of the main contributions was produced in Isard (1949).

While in Civilian Public Service he also continued his research on the implications of the new transportation technology, namely aircraft, for location of industry and urban development, a subject of considerable national interest at that time.

At the end of his Civilian Public Service work in 1946, he was granted a post-doctoral fellowship of the Social Science Research Council. He returned to Harvard University to continue his basic location studies. With the encouragement of Professor Abbot Payson Usher he studied location factors in the iron and steel industry. Recall that in the late 1940's that industry was still considered the most basic industry for economic development, being well known for its major multiplier effect. Further, it was viewed as immensely important for military operations and national security, although aircraft development and associated advance in electronics was soon to replace iron and steel as the most critical production activity for military ascendancy and supremacy. For example, with regard to iron and steel's potential economic impact, there was a major effort to stem the decline of the New England economy with the construction at New London or Fall River of an integrated iron and steel works. Because of the perceived importance of iron and steel development, there was widespread interest in the three articles on the location of the iron and steel industry published (Isard, 1948; Isard and Capron, 1949; Isard and Cumberland, 1950). Once again the relevance of location analysis became widely appreciated in non-academic circles, this time among industrialists and business leaders.

Further, during the time he spent in Civilian Public Service, the atomic bomb was dropped. The reaction to this massively destructive weapon was widespread and intense. To counter the unimaginable loss of life and physical scourge inflicted by a single bomb, a number of nuclear physicists (including Nobelists) glowingly pointed up beneficial possibilities—extremely cheap power for all nations of the world and the consequent economic flowering of all the poor regions. The son of immigrant parents, recently versed in the Viner tradition of cost analysis, tested the off-the-cuff cost estimations of the distinguished physicists. After digging deeply into the technical literature, he came up with kilowatt hour costs much higher than those hypothesized by physicists and others. The basic reason for this conclusion would be the extremely large fixed costs from the need for huge capital investments in facilities to guard against radiation hazards—some necessarily to be met by government subsidization. Together with a colleague, Vincent Whitney, he extensively wrote articles in diverse publications as the *New Republic,* the *Harvard Business Review*, *Yale Review*, the *American Scholar*, the *Bulletin of the Atomic Scientists*, as well as technical journals such as the *Quarterly Journal of Economics* and the *Review of Economic Statistics*. These seriously questioned the positive economic impact of atomic power. The production of such power would involve high costs. Poverty-stricken regions would not prosper. Very little new industry, if any, would come to be located in them. Only in major industrialized countries, with tremendous demand for power large enough to make feasible extremely large plants, could atomic power be competitive, if indeed it did not require burdensome subsidization by these countries. The extensive publication and participation by this son at major national conferences on atomic energy brought the relevance of the subject of industrial location and regional development to the attention of a diverse lay and professional public.[5] While the Anglo-Saxon economists, except for Alvin Hansen and Abbott P. Usher and some of their students, continued to ignore location analysis, the rest of the knowledgeable world became well aware of the significance of this subject. This was still more the case when in 1948-49 that son became Associate Director of the Teaching Institute of Economics at American University, Washington, D.C. whose primary concern was with research and teaching of the implications of atomic energy.

Recall that in the pre-World War II years, the input-output approach to economic analysis developed by Wassily W. Leontief had been steadfastly ignored. However, during the U.S. participation in the war, the problem of bottlenecks in production became critical. Duane Evans and Marvin Hoffenberg, in charge of U.S. government research on production bottlenecks, exposed the extremely important use of input-output for anticipating such (Evans and Hoffenberg, 1947). Thus, with the demonstration of its critical war value, and the consequent awakening of economists to its major potential for analysis of peacetime development, input-output became a rage among economists. In the 1950s and '60s input-output along with the earlier Keynesian Revolution constituted two major thrusts in economic analysis.

As a consequence of the new promising applications of input-output, Leontief found himself with abundant research funds. In looking for fruitful directions in which to use these funds, he hit upon regional problems. Some regions, especially New England where Harvard University was located, were facing difficulties arising from the cessation of military production. In his search for young scholars trained to handle both mathematics and location/regional analysis he found very few, if more than one, to choose from. Thus, to move ahead, he had to consider that son for assisting him in his regional work—in particular for the development of Leontief's balanced regional input-output model. That son, however, was not willing to return to Harvard, unless he was given a course to teach in the Economics Department. There was none for which the Harvard faculty considered him qualified to teach. However, Leontief persuaded the faculty to set up a new course for him, one on location theory. Such a course was taught at only one leading university in U.S., namely the University of Michigan. It was offered there because on its staff was a fine scholar Edgar M. Hoover, who had written a notable book

---

[5] At one memorable occasion, as a physical lightweight, he was squeezed in on a platform between two hefty heavyweights, one Norbert Weiner, the mathematical genius from MIT, and the other, Edward Condon, a distinguished physicist, head of the U.S. National Bureau of Standards. He made clear that there could be only highly limited economic implications because of the costliness of nuclear power. No flowering of undeveloped regions with new locations of major industry would occur.

on the Location of the Shoe and Leather Industry. The Harvard faculty was willing to accede to Leontief's request. It expected that few, if any, students would elect to take that course, and that it would soon be dropped because of very limited, if any, attendance. In June 1949, that son joined the Harvard project.

There was one more factor contributing to the development of regional science. In academic year 1949-50 the United States was still engaged in a major shift to peace time operations. Many of the Harvard professors were busy consulting in Washington, D.C., where their advice was urgently needed. To the bright wartime veterans coming to Harvard for graduate study and anxious to get along with their careers, Harvard, with a number of its professors not available for consultation and advice, was indeed a cold place. These students learned of the new location course, sampled it, found it interesting, and more important found its instructor a person with whom it was easy to arrange appointments to discuss their problems, often those of adapting to civilian life. Also, in the approved course, this instructor introduced input-output analysis, for the first time to be taught at Harvard. (In typical modest Continental European fashion, Leontief did not teach his major contribution, input-output analysis, in any of his courses.) To the impatient veterans (graduate students) tired of poring through theoretical volumes and wanting practical knowledge leading to jobs, the translation of the highly mathematical Walrasian general equilibrium system into what they considered to be a practical set of numbers was extremely exciting, as it was also to the instructor. Thus as input-output became more and more recognized as an effective tool of analysis, more and more students came to attend that instructor's course. This growing attendance and interest in the topics covered in the one-semester course led to the reorganization of it as a year-long course on both location analysis *and* regional development. The course also made it possible for the instructor to become acquainted with students and to select outstanding ones with whom to conduct joint research. Once he had worked successfully on such research leading to jointly authored articles (where the graduate students had contributed equally if not more to the quality of the output), this outcome became known. Hence, in later years when regional science was established as a field, this led to a continuous inflow of top-notch students. And with the passage of time, these students increasingly contributed more than the instructor to journal articles.

Thus by 1949 that entrepreneurial type personality had acquired recognition as a locational theorist and analyst in the academic community and several significant sectors of the business world. But location analysis and problems are only one basic aspect of regional analysis. Further for the emergence of regional science and the Regional Science Association there still was absent both resources and ties with a spurt of new creative thinking in an established and conventional discipline. To these needs we now turn.

2.2 Growth of Interest in Regional Problems, Informal Meetings of Regional Researchers and Promotion Efforts

Recall that Leontief was flush with research funds and always had abundant secretarial staff. Leontief easily agreed to allow me,[6] his newly acquired research associate, fully to exploit that resource for arranging conferences, especially since these conferences would further point up the value of both general and regional input-output analysis. Thus, in one way of thinking it may be said that the birth of regional science can be attributed to the first successful conference on location and regional matters that the instructor organized. As will be noted below, that birth was on December 29, 1950.

Before joining the Leontief project in June 1949, I had returned to Harvard University to complete my studies under the Social Science Research Council sponsorship, interrupted by the war. While engaged in such, I was active in efforts to develop the study of location and regional problems and trying to arrange whenever possible gatherings of scholars to promote relevant analysis. At the annual conventions of the American Economics Association (AEA) I sought to have sessions devoted to these problems and analysis. I wrote Presidents of the Association on this matter: (1) to Joseph A. Schumpeter in 1948, who was interested primarily in business and trade cycles and not on regional analysis and development—however, Schumpeter did set up a session on cycles experienced by regions and on the interregional transmission of them; (2) to Howard Ellis in 1949 with no success; (3) to Frank H. Knight in 1950 also with no results. However, Philip Neff and I requested President Knight to allow regional scholars to announce and hold an unsponsored meeting at a convenient time and place during the 1950 December convention. Such permission was granted.

In these unsuccessful efforts at formally programming location and regional development sessions at AEA conventions, I worked with scholars such as Joseph J. Spengler, Yale Brozen, and Stefan Robock. In my communication and interaction with Joseph J.Spengler, a population expert, I had occasion to bring to the attention of Spengler and later on of Anglo-Saxon economists, the works of the German location scholars.[7] In a letter to Spengler I wrote that I had occasion to reread many of my notes on the German locational writings. Some of them, such as those of Roscher, Schäffle, Engländer, Ritschl and the less known parts of Weber (1909)

---

[6] From here on, I shall no longer refer to myself (W. Isard) as the son of immigrant parents.

[7] For some of these works, see Isard (1949). This article also sharply criticized Hicks, Mosak and Lange, to name a few, for ignoring space as a basic factor in their general analyses in the 1940s. It also pointed to the significant improvement possible by replacing the concept of pure competition with that of monopolistic competition given that the latter can incorporate more of reality, especially spatial effects.

(his Chapter VII and his *Grundriss der Sozialökonomik*) suggest an approach to study of the long-run tendencies of population location, which might be very fruitful to explore. As you know, aside from Hoover's two excellent books, there is little worthwhile Anglo-Saxon literature on locational analysis. And Hoover's approach is one of partial equilibrium analysis. Emphasis on the approach of the German Historical School, I think, is more suitable for the analysis of the changing spatial distribution of population. A paper carrying forward this approach would not only be informative but could develop useful analytical techniques.

Incidentally, a study of the determinants of population location is not unrelated to development in state regulation of and participation in economic activity. The decreasing material orientation of industry and the increasing market and labor orientation of industry can be put in German locational terms. The thesis that non-economic factors are becoming relatively more significant in the location of population can reflect, I believe, a decreasing economic penalty when the state interferes with or redirects economic developments for the welfare of its people. I suspect that a good deal of the British experience with state regulation and participation, especially with reference to the iron and steel industry where social costs in the movement of population have weighed heavily, can be interpreted more meaningfully in this context.

This emphasis on the contributions of the German Historical School and the German locational writings led to a major overlap of regional research and later regional science with one more social science field, namely demography and population studies. This overlap continues today in the migration and related studies of regional scientists, and is manifest in the first two chapters of Isard (1960).

Returning to the problem of obtaining greater appreciation of the value of regional research, the set of scholars already mentioned and others initiated major efforts to have recognition of the significance of regional studies by the Social Science Research Council and financial support of their research by Foundations (Carnegie, Rockefeller and others).

One effort by Yale Brozen and Isard resulted in a proposal for research on the impact of technological change with especial attention to effects on industry location and regional development. Another effort, by Philip Neff and Isard, in a letter of November 4, 1950 to the Social Science Research Council, was a proposal that the Council establish a committee on regional economic studies with a set of objectives and tasks that were to be thoroughly discussed and revised at the forthcoming December 1950 meeting of regional researchers.

Given the permission to hold an unsponsored meeting at the time (1950) and place of the convention of the American Economics Association, I wrote numerous letters to interested scholars inviting their attendance. The minutes of the December 29, 1950 meeting, which were distributed with the January 9, 1951 newsletter, are presented below. They well illustrate the diversity of problems, motives and interests that existed among regional economists.

Newsletter, January 9, 1951. Littauer 307, Cambridge, Mass.

Enclosed is a condensation of the proceedings of the Regional Economic Research meeting held at the Palmer House, Chicago, December 29, 1950. I regret that we were not able to take down all the comments of everyone present. If any participant's viewpoint is misinterpreted, please let me know so that I may make corrections.

May I gently remind you to send in your opinions and suggestions relating to critical areas for future regional research and to the functions of the proposed Committee on Regional Economic Studies—if you have not already done so. Also, let me know whether or not you wish to participate in a summer research seminar for a week or as a member of the core group, if the Social Science Research Council approves this project.

You may recall that many emphasized the basic need for an increase in the supply of regional data. Right now I am preparing a memorandum on ways in which 1950, 1947 and other census data might profitably be cross-classified to throw light on the locational structure of population and industry in the United States. Here lies an opportunity to increase the supply of data for regional research by presenting existing data in more usable form. May I have your suggestions as to desirable cross-classifications by February 1?

With best wishes for the New Year, Sincerely, Walter Isard.

The memorandum that was prepared is contained in Appendix A.

Proceedings of the Regional Economic Research, Meeting Held at the Palmer House, Chicago, Illinois on Friday, December 29, 1950

The participants were:

- Scott Keyes, Pennsylvania State College, State College, Penna.
- E.T. Grether, University of California, Berkeley, California
- Glenn E. McLaughlin, National Security Resources Board, Washington, D.C.
- Philip Neff, U.C.L.A., Los Angeles, Calif.
- Nathaniel Wollman, University of New Mexico, Albuquerque, N.M.
- Milburn L. Forth, Commonwealth Edison Co., Chicago, Illinois
- Morris E. Garnsey, University of Colorado, Boulder, Colo.
- Samuel H. Thompson, B.L.S., Dept. of Labor, Chicago, Illinois

- Maurice E. Moore, B.L.S., Department of Labor, Chicago, Illinois
- Marvin H. W. Derner, B.L.S., Dept. of Labor, Chicago, Illinois
- Stefan H. Robock, T.V.A., Knoxville, Tenn.
- Hiram S. Davis, University of Pennsylvania, Philadelphia, Penna.
- James N. Morgan, University of Michigan, Ann Arbor, Mich.
- J.P. Watson, University of Pittsburgh, Pittsburgh, Penna.
- James C. Nelson, Washington State College, Pullman, Wash.
- Robert L. Steiner, 3443 Stettinius Ave., Cincinnati 8, Ohio
- Harry S. Schwartz, Federal Reserve Bank, San Francisco, Calif.
- Fred Deming, Federal Reserve Bank, St. Louis, Mo.
- Reinhold P. Wolff, University of Miami, Miami, Fla.
- Coleman Woodbury, Urban Redevelopment Study, Chicago, Illinois
- Joseph L. Fisher, Council of Economic Advisers, Washington, D.C.
- Werner Hochwald, Tulane University, New Orleans, La.
- Jacob A. Stockfisch, Federal Reserve Bank, Kansas City, Kansas
- Byron M. Williams, U.C.L.A., Los Angeles, Calif.
- John Blackmore, T.V.A., Knoxville, Tenn
- Walter Isard, Harvard University, Cambridge, Mass. (Chairman)
- Leon Moses, Northwestern University, Evanston, Ill. (recording secretary)

Chairman:

Purpose of meeting:

1. to exchange ideas on what should be done in regional research,
2. to discuss a proposal made to the Social Science Research Council on the formation of a committee on regional economic studies.

History of proposal was sketched and letter of November 4, 1950 to Pendleton Herring from Neff and Isard was read.

Meeting was opened for discussion.

Neff:

Suggested that in view of the present emergency the committee might prove most helpful if it were to formulate its program in terms of both short and long run goals.

Grether:

Strongly urged an interdisciplinary attack upon regional problems, one that would involve the cooperation of political scientists, sociologists, geographers, engineers, and other scientists. Mentioned the West as an area which has problems national as well as regional in scope. The legal and political aspects of these problems (in terms of national legislation) cannot be set aside.

Thought it unwise to overemphasize the problems of short-run mobilization.

Wollman:

Mentioned the importance of problems of industrial decentralization. Associated with these are both short-run and long-run phenomena, each of which require evaluation and study.

Garnsey:

Suggested that we determine the limits to industrial decentralization.

McLaughlin:

Asked whether development in one region takes place at the expense of others. How can plans be put together to form a consistent whole? Strongly emphasized that to answer such a question one must develop new concepts and techniques.

Neff and Chairman:

Stated that at the start the committee ought not to be inter-disciplinary in character but ought to consist solely of economists. And that only after the committee had gained momentum would it be desirable to bring in members of the other social sciences. This might avoid the problem of being bogged down at the start with too many questions.

Grether:

Partially conceded the point but urged that at least a political scientist and a geographer be included on a committee.

Woodbury:

Stated that it was important to bring the other social scientists in from the start because of the importance of the group's growing together.

Fisher:

Said that there were many unrelated groups doing research on particular areas. The chief problem is to draw them all together by some overall theorizing and philosophizing. As a point of attack, suggested definition of such concepts as (a) lag, (b) balance, (c) development, (d) decay.

Heartily endorsed the task stated in the letter of November 4, "The establishment of a set of clearly defined, standard, analytical concepts ..."

Hochwald:

Pointed out the limitations of economic analysis as such and felt that the concept of the region as a problem area, because it requires the contributions of the various social sciences, was a very useful tool for integrating these sciences.

Wollman:

Turned to another possible function of a committee on regional economic studies. Stated that much data which is collectible is not in fact being collected. The committee could foster the collection of such data.

Garnsey:

Pointed out the need for a general theory and philosophy to determine which data should be collected. Tentatively suggested collection of data on gross regional product, regional flows, and regional investment. All these data would be highly relevant to short-run defense analysis.

Various Individuals:

Suggested the Bureau of Labor Statistics, the Securities Exchange Commission, the Department of Commerce, the Federal Reserve Banks, the Interstate Commerce Commission, the Treasury and the Bureau of the Census as data collecting agencies which should be advised in presenting and collecting more data for regional analysis.

Steiner:

Suggested that after a committee was well organized and underway, a book be edited, containing contributions by regional economic researchers on various regional problems. This would stimulate interest and be extremely helpful to people outside the economics profession, especially to city and regional planners.

Referred to the possibility of a summer seminar to evaluate the various ideas arising in the discussion, to shape and synthesize them in order that they could form the basis for chapters in the proposed book.

Schwartz:

Returned to the problem of needed concepts. What does it take to get employment in an area? It is not sufficient to note only the employment in the new industry. Analysis must proceed farther to determine the amount of employment which the workers in the new industry generate. New concepts, new coefficients are needed to quantify these indirect effects.

As an example, a region may be declining because it has lost its advantage with respect to certain lines of activity. In other lines it may have comparative advantage, and these alone should be encouraged. But the resulting employment will be more than the employment in these lines of comparative advantage alone. To tell us how much more, we need new concepts.

All problems involve political science, sociology, engineering as well as economics. Pointed to the problem of balance as an illustration.

McLaughlin:

But what is balance? Can it be defined? This is associated with the tough problem of determining cost and benefits of various public works programs. Stressed necessity for interdisciplinary attack on this problem.

Fisher:

General assent. Sociology and psychology are required in order to give meaning to concepts of benefits and costs, despite the fact that you can't quantify them in monetary terms.

Robock:

Suggested that we set up a working committee to sort out ideas, to determine valid approaches.

Johnson:

Said we lack both a theory of regional development, and the data with which to work.

Neff:

There is some helpful general theory, such as Ohlin's. We need not concern ourselves at the beginning with a general theory of development. But we need data to test theories we evolve.

Nelson:

What kind of data do we need? Data on flows seem most important in order to cast light on the relations of one region to another. Much material collected, but not put in usable form for presentation.

Johnson:

We need continuing data collection, not for one point of time. We have many spot studies for different regions at different points of time. Lack of continuity detracts considerably from their value.

Thompson:

Stated that even if economists have not developed a general theory, their advice on the construction of roads, public works, etc. should be made available to engineers and others who frequently carry on their projects without thinking in terms of regional economics.

Chairman:

Three basic ideas thus far expressed:

1. interdisciplinary attack
2. need for new concepts and techniques which relate each region to every other region and to the nation—for frameworks for making consistent regional projections.
3. increase in the supply of available data

General discussion followed, primarily on the first basic idea. It was finally agreed that a committee composed of a core of active regional economists with perhaps a few other intensely interested social scientists would be best at the beginning. Later, as the committee gained momentum, additional non-economists should be added.

The possibility of an interuniversity summer research seminar was discussed generally. The suggestion was made that in view of the fact that many individuals cannot leave their work for six or seven weeks, it might be more desirable to have many individuals participating in a seminar running one or two weeks.

Objection was taken on the ground that not much could be accomplished in one week or two, if all were to devote only one or two weeks to the seminar.

A compromise suggestion was made and generally accepted as most desirable. The seminar could consist of two groups of individuals. One, of researchers who would be attacking basic problems for a period of six to seven weeks. This would be a core group. Another, of individuals who, a few at a time, might spend approximately a week participating in the seminar, discussing their particular regional problems with members of the core group.

A conference of regional researchers to run for three days each year was suggested. After much discussion it seemed desirable to have at the time of each annual A.E.A. Convention, at least one meeting, scheduled but not public, devoted to regional problems.

The problem of a working committee was taken up. After much discussion a motion was passed to the effect that the presiding chairman also be chairman of the working committee and that he appoint the members of such committee, provided they be chosen to represent the various geographic areas. The committee was instructed to report to those assembled on the progress of its work, which is to digest the ideas expressed at this meeting and to make recommendations for action.

Expressed active interest, but unable to attend:

- Edgar M. Hoover, Council of Economic Advisers, Washington, D.C.
- Dewey Daane, Federal Reserve Bank, Richmond, Virginia
- Victor Roterus, Department of Commerce, Washington, D.C.
- Simon Kuznets, University of Pennsylvania, Philadelphia, Penna.
- Harold Williamson, Northwestern University, Evanston, Illinois

- Rutledge Vining, University of Virginia, Charlottesville, Va.
- Keith Johnson, Federal Reserve Bank, Dallas, Texas
- George Garvy, Federal Reserve Bank, New York, N.Y.
- Werner Z. Hirsch, University of California, Berkeley, Calif.
- Paul Zeis, Dept. of Commerce, Washington, D.C.
- Joseph J. Spengler, Duke University, Durham, North Carolina
- Calvin Hoover, Duke University, Durham, North Carolina
- Arthur Upgren, University of Minnesota, Minn.
- James Marti, University of Kentucky, Lexington, Ky.
- Guy Freutel, Harvard University, Cambridge, Mass.
- Dexter Keezer, McGraw-Hill Publishing Co., New York, N.Y.

Working Committee

- Morris E. Garnsey, University of Colorado, Boulder, Col.
- E.T. Grether*, University of California, Berkeley, Cal.
- Werner Hochwald, Washington University, St. Louis, Mo.
- Glenn E. McLaughlin, National Security Resources Board, Washington, D.C.
- Philip Neff, University of California, Los Angeles, Cal.
- Stefan H. Robock, Tennessee Valley Authority, Knoxville, Tenn.
- Harold Williamson*, Northwestern University, Evanston, Ill.
- Walter Isard, Harvard University, Cambridge, Mass. (Chairman)

  * Appointed, subject to his approval

As already mentioned, the meeting was extremely successful in bringing together a diverse group of scholars. The high level of participation and quality of discussion at the meeting, along with the availability of free secretarial assistance from the Harvard input-output project, almost made inevitable subsequent meetings at the time and place of later American Economics Association conventions.

After the stimulating December 29, 1950 informal meeting, intensified effort continued to be made by interested regional scholars to obtain financial and other support from foundations and the Social Science Research Council. However, proposals initiated and made both before and after that meeting were turned down, even after the foundations and the Social Science Research Council were informed about that meeting's success. In particular, a very carefully constructed proposal to establish a Committee on Regional Economic Studies was turned down by the above Council. The proposal is embodied in the following letter.

Mr. Pendleton Herring, Social Science Research Council, New York, New York January 9, 1951

Dear Mr. Herring:

We wish to report on the outcome of the meeting held with our colleagues this Christmas. You may have heard by now that the response was very gratifying, and that the interest in improving the quality of regional research was intense.

Enclosed you will find a condensation of the discussion at the meeting, and lists of

1. participants
2. those expressing a strong interest but unable to attend, and
3. members of the working committee.

As a result of our discussion, the program of a committee on regional economic studies tentatively proposed in our letter of November 4 might be revised to read as follows:

The general objectives, we think, should be to foster continuing discussion and the exchange of ideas and information among individuals and organized groups engaged in regional research in the United States, to appraise the present status of regional research, and to delineate critical areas within which additional research might make the greatest contribution. At the outset, the committee on regional economic studies might concentrate on the following specific tasks:

1. The establishment of a set of clearly defined, standard analytical concepts in order to permit comparison of research findings of the various individuals currently at work in this field. We feel that this would significantly accelerate the accumulation of economic knowledge. Attention should be given to the development of new concepts more significant than the old and consistent with recent advances in general economic theory and with the changing set of interregional relations.

   Many problems (raised at our Christmas meeting) require new concepts and techniques: for example, the continuing problem of measuring benefits and costs of public improvements in any region; the definition and measurement of regional lag, of regional balance, of regional income, of gross regional product and investment, of regional growth and decay; the evaluation of trends in regional structures; the determination of the economic limits to industrial decentralization; the inclusion of a metropolitan focus in the categories of regions; the measurement of the effect of changes in industrial composition upon regional employment multipliers and the need to reorient old concepts, such as primary, secondary, and tertiary activities to existing interregional flows of goods and services.

2. The examination of the explicit and implicit theoretical structures and frames of reference of current research, particularly with respect to relationships between regions as such, between various regions and the American economy as a whole, and between these and the international economy. This should be done with a view to reconciling and synthesizing the various approaches where possible in order that the findings of many research studies may fit together more meaningfully within a common framework.

   This examination might go far toward meeting the need (which many participants stressed) to evaluate a given region's development in the light of development in others. Are the various regional programs consistent with one another and with development of the national economy? Obviously, the answer to this question requires an analysis of the assumptions underlying each region's plans and of the nature of interregional relationships.

   The performance of such a task might also greatly assist in the development of techniques (for which participants felt a strong need) which not only could translate national projections into a set of regional projections, but also could build up a national projection from a set of mutually consistent regional projections.

3. The promotion of an increase in the supply of data useful to regional analysis. This can be accomplished through advising data collecting agencies such as the Bureau of Labor Statistics, the Securities Exchange Commission, the Federal Reserve Banks, the Department of Commerce, the Interstate Commerce Commission, the Treasury, and the Bureau of the Census: (a) on data vital to regional research which is collectible but is in fact not being collected. (For example, the Census of Manufactures ought to separate transport costs both on inputs and outputs from total production cost to cast light on the role of transportation in the location of in-

dustry); (b) on ways in which collected data should be classified in order to be most useful for regional research; and (c) on existing imbalances in data collection which ought to be corrected. (For example, for the study of flow phenomena the abundance of data on railroad shipments contrasts sharply with the paucity of data on truck shipments.)

We also had the opportunity to discuss the desirability of an interuniversity summer research seminar on regional problems. Many individuals are interested in participating, in sharing ideas and gaining insights into the approaches of others, in subjecting their own methods of analysis to sharp criticism. However, few are in a position to devote six to seven weeks to such an undertaking. It was generally agreed that the value of the seminar might be maximized, given the budget limitations, if the seminar were to consist of two groups of individuals. One, of researchers (perhaps four) who would be attacking basic problems for a period of six to seven weeks (this would be the core group). Another, of individuals, who, a few at a time, might spend approximately a week on the project, discussing their particular regional problems with members of the core group and acquiring fresh viewpoints. Thus far, Moses, Isard, and Hirsch have expressed a desire to participate as members of the core group. Neff, Robock, Stockfish, Fisher, Williams, Garnsey, Johnson and Grether are among those wishing to devote a week to the seminar (no attempt was made to poll those desiring to participate for a week.) It is anticipated that many more will be interested. Since excellent library facilities will be required, the seminar should be held either at the University of Chicago or Harvard University. The former seems preferable because of its central location.

We very much hope that the proposed program for a committee on regional economic studies is stated in sufficient detail to allow the Problems and Policy Committee to consider at its January meeting the formal support of such a committee. Should the Problems and Policy Committee also approve an interuniversity summer research seminar on regional problems, it appears that much of the expense of the first year's operation of the committee on regional economic studies could be avoided by (1) convening during the summer at the university where the seminar is located and (2) encouraging members of the seminar core group to initiate studies related to the proposed tasks of the committee.

With every best wish for the New Year,
Sincerely, Walter Isard and Philip Neff

Herring's response of February 13, 1951 to the January 9, 1951 letter was:

> Dear Dr. Isard,
>
> I am sorry that it is necessary to report that our Committee on Problems and Policy last Saturday concluded not to go ahead at this time with the establishment of a committee on regional economic studies and that it also decided not to approve your pending summer seminar proposal.
>
> With respect to the proposal for a committee, the reasons were essentially those which we discussed when you were in the office here. This decision might be reconsidered sometime later, but certainly not until a good many of our current responsibilities are further advanced than is the case at the present time.
>
> The summer research seminar proposal was turned down principally because of your suggestion that the number of full time participants be limited to three or four and that a number of others be invited to participate only for a week or so. This would be in decided conflict with the promise on which the funds for the seminar program were obtained by the Council, namely, that the funds would be used to assist a number of able individuals in devoting a summer to research instead of having to spend the summer in earning enough money through non-research activities to balance individual or family budgets. Your memorandum of course listed only two commitments for full time participation and one additional name is uncertain.
>
> In view of the action taken I am returning the correspondence which you kindly loaned to me last November.
>
> Sincerely yours, Pendleton Herring

Thus, with the rejection by the Social Science Research Council and foundations of proposals that were submitted, the only avenue for the development of locational and regional economics was through continuing to hold informal meetings when and wherever a fruitful occasion presented itself. These would be at conventions of the American Economics Association and others like the American Sociological Society, the American Institute of Planners, the Association of American Geographers, the American Political Science Association, the American Statistical Association and the American Association for the Advancement of Science, particularly where there were overlaps of interest among regional researchers and other social scientists, planners and engineers.

Although we were yet to be successful in obtaining funds for our projects, although we were still subject to the negative Anglo-Saxon bias, we had much momentum and enthusiasm for regional research. Given that we had access to necessary secretarial assistance from the Harvard input-output project, many of us judged that with concerted effort we could move forward with informal meetings; and some scholars concluded that it was a waste of time, or at least an unproductive use of our time, to deal with foundations. The above letter of January 9, 1951 to Pendleton Herring and the minutes of the December 29, 1950 meetings clearly testify to the fact that the group had a broad but well-defined set of interests which were complementary and which could be fruitfully coordinated. Thus the regional researcher at the Harvard input-output project was encouraged to persist, and aggressively so, in organizing meetings and searching for research support for regional projects. The fighting spirit of the group is indicated in the following letter of March 24, 1951.

Dear Dean Grether,

It was good to have your letter of February 28th. I have now had a chance to correspond with members of the working committee and want to report certain reactions.

First, it is clear that none of us wants to give up the fight. All have urged continuing efforts despite our initial reverses. As you can see from the enclosed letter to Mr. Herring I am hoping to be able to take up with him again the issue of Council support. Further, we all agree that meetings at regional as well as national association conventions are desirable. Formal and informal sessions have been arranged for the Midwest Economics Association convention to be held in April. I have spoken to John H. Williams and written him at his request re: regional sessions at the A.E.A. meetings and have spoken to W.W. Leontief, program chairman for the Econometric Society, on the same matter. I shall be taking steps to arrange sessions at other conventions.

In addition, there is strong feeling that we ought to approach Ford Foundation for support of a conference. It is on this matter that I should like to turn to you for leadership. You indicated at our Christmas meeting that, upon prodding, you could be induced to approach the Ford Foundation directly, or indirectly through close associates at the Universities of California, North Carolina, etc. And in your letter of February 28 you state that you will keep your "eyes and ears open for possible sources of funds". I do not know how far you have gone in seeking financial support, but if you have not approached the Ford Foundation, may we rely upon you to do so as soon as is convenient? In hopes that we may, I am enclosing relevant comments by Phil Neff which may be helpful.

Shall keep you informed of further developments. Am planning to send out minutes of our Midwest meeting.

Sincerely yours, Walter Isard

A proposal that was prepared by Dean Grether and submitted to the Ford Foundation is contained in Appendix B. It was turned down.

Despite the continuing negative responses to our initiatives[8] the organization of meetings at opportune places and times continued. Below is a report of an informal meeting that was able to be arranged at the Midwest Economics Association convention on April 20, 1951. This report continues the live discussion of the December 29, 1950 meeting. It concentrates at depth on the issues of appropriate data requirements for regional research and collection problems—some of which are still with us today.

Proceedings (Selected) of the Regional Economic Research Meeting Held at Hotel Pfister, Milwaukee, Wisconsin, on Friday, April 20, 1951

The participants were:

- Werner Hochwald, Tulane University, New Orleans, La.
- Philip M. Faucett, Jr., Federal Reserve Bank, Chicago, Illinois
- William M. Capron, University of Illinois, Urbana, Ill.
- Charles G. Wright, Federal Reserve Bank, Chicago, Ill.
- Robert L. Bornholdt, Federal Reserve Bank, St. Louis, Mo.
- Clarence W. Tow, Federal Reserve Bank, Kansas City, Kansas
- W. Kowisto, Lake Forest College, Lake Forest, Ill.
- Maurice E. Moore, B.L.S., Dept. of Labor, Chicago, Ill.
- Edgar Z. Palmer, University of Nebraska, Lincoln, Neb.
- Theodore F. Marburg, Hamline University, Minneapolis, Minn.
- Samuel L.M. Loescher, Indiana University, Bloomington, Ind.
- Harold F. Williamson, Northwestern University, Evanston, Ill., chair, local arrangements
- Gilbert M.Mellin, Tulane University, New Orleans, La.
- Leo W. Sweeney, State University of Iowa, Iowa City, Ia.
- Earle W. Orr, Jr., State University of Iowa, Iowa City, Ia.
- R.D. Pashek, University of Illinois, Urbana, Ill.

[8] Another negative response was from the National Research Council. With the support of William G. Friedrich an effort had been made to obtain financial support for regional statistical studies.

- J.A. Buttrick, Northwestern University, Evanston, Ill.
- F.M. Boddy, University of Minnesota, Minneapolis, Minn.
- E.E. Hagen, University of Illinois, Urbana, Ill.
- Professor Ryan, University of Wichita, Wichita, Kansas
- Walter Isard, Harvard University, Cambridge, Mass. (Chairman)
- Guy S. Freutel, Washington University, St. Louis, Mo. (Recording Secretary)

Chairman:

Purpose of informal meeting:

1. to bring regional researchers up to date on developments in regional research.
2. to discuss specifically the problem of increasing the supply of regional data.

General remarks were made on the history of our effort to improve regional research. The proposal for support of a committee on regional economic studies by the Social Science Research Council was reviewed. The full statement of the third specific objective in that proposal, specifically "the promotion of an increase in the supply of data useful to regional analyses" was read.

The meeting was then opened for discussion.

Moore:

Remarked that he had been contacted by a representative of the Society for the Advancement of Management, who stated that many persons in industry were interested in contributing realistic information on a regional basis, and in turn receiving advice on ways of attacking problems confronting them. He wished to establish better communication with regional researchers.

Chairman:

Suggested follow up. Specific manner to be discussed after session.

Hochwald:

Should consider the relative role of different sources of data. The role of census data is declining relatively, for the nation as well as the region. Other sources of data are increasing. Therefore, census data are and should be increasingly used as benchmark data.

Stressed the importance and greater need for sample studies (e.g. intercensal) for obtaining at a relatively low cost current regional data consistent with census benchmark data. This, to illuminate continuously regional changes.

Also many administrative records exist which at low expense could be kept in such a way as to be useful for regional analysis. Though the O.A.S.I. data have been made available, there are many data which have not. E.g. state tax returns which could provide a mine of information on local income patterns. Also, it would be highly desirable to have federal fiscal expenditures specified regionally. Suggested that this group could make a contribution by urging that administrative records be kept in such a way as to be available on a regional basis and meaningfully related to census benchmark data. This would afford a way of adjusting benchmark data for current use.

Chairman:

Asked for reaction to these suggestions. Approval was general.

Capron:

For diverse administrative and research groups, the unit geographic areas for data collection are different. At the same time the geographic areas demarcated as regions are also different so that frequently regional data are not comparable. Pointed out, for example, that metropolitan analysis requires data which unfortunately are collected by various agencies on a state basis only. Could this group agree on the minimum size unit for data collection to insure that data can be aggregated in the various ways which are required by different regional studies?

Boddy:

How to define a region? Isn't it necessary to define a region first before you can specify the most desirable minimum size unit?

Hochwald:

But how can you define the boundaries (in contrast to focal points) of regions if you don't have the data for small areas to reflect the shading off of relations.

Sweeney:

Concurred. Cannot define uniquely a region. But can usually fit county data into any desirable design of regions. Raised the question of obtaining data on a county basis.

Boddy:

Questioned feasibility of obtaining certain data on a county basis.

Hagen:

Warned against danger of imbalance in regional research efforts. Ideally we want all kinds of data by households. But discouraged getting all kinds of data if such would divert energies at improving and developing concepts and techniques for regional analysis. On the other hand, we should encourage and facilitate the collection of data which would not involve effort on our part. This data in general will only be collected by administrative agencies where useful to them. Thus, if we are to obtain certain data for regional analysis, we must point out their value for other agencies and purposes. Reiterated that where county data are available at little expenditure of effort, they should be gathered. Hesitated at putting too much effort into the collection of data alone.

Chairman:

Can we narrow down the range of minimum size units for which data may be desirable? Should we seek data on units smaller than counties? Should we consider parts of counties as the recent census does in distinguishing between urban and non-urban areas? Should effort at this type of breakdown have any priority at all?

Hochwald:

This is a problem of relating inputs to results. We already have data for certain areas smaller than counties, such as population and census of business data for census tracts and certain blocks in urban regions. These obviously should be utilized.

Asked how useful data by "economic areas", as defined by the 1950 Census, might be for regional analysis? Since sizes of counties differ widely among states, use of "economic areas" defined in terms of homogeneity of economic activity might permit more consistent analysis.

Chairman:

Suggested that this last question be tabled for discussion at some later meeting when we all will be familiar with this new concept and when the complete set of data by economic areas is available.

Returned to basic question. Can we take a positive stand and make this resolution:

In general we should seek considerable data on a county level, except in those instances where costs may be excessive. In general, too, we should not seek data on a finer breakdown, except again in isolated instances where almost absolutely essential for research.

This resolution was adopted subject to the qualification that there be periodic reexamination of the allocation of effort to data collection on one hand, and analysis on the other.

Capron:

Our discussion implies that data on a county level would be sufficient for all designs and regions and types of analysis.

Chairman:

Such data would be sufficient for most regional researchers except perhaps human ecologists. To collect data profitably for a finer level than the county would require considerable additional and more refined analysis. If we were in fact to seek such data, we would have imbalance in the sense that our conceptual framework would be lagging far behind our empirical work.

Capron:

This implies also that state data alone are inadequate.

Chairman:

Yes. In practically all regional studies, state and regional boundaries are not coincident. In regional input-output analysis for example, we run into difficult problems because of this lack of coincidence.

Palmer:

Regional research studies at Nebraska have successfully used sampling techniques on rural counties, where county data are not available, in combination with the complete set of data for urban counties. This is an effective method of economizing in effort at data collection.

Wright:

Discussed certain problems in sampling. In particular, he pointed out that frequently sampling studies use different groupings of states to form approximately the same region, as dictated by the different purposes of these studies. Urged seeking more agreement on selection of regional boundaries. Suggested that we explore the possibilities of obtaining regional data from the samples of the Survey Research Center, especially where locality of each sample item is known.

Chairman:

Given our resolution, let us go on to another question. What types of data should be sought? Do we all agree that regional income data would be desirable to have?

Hochwald:

This question raises the problem of defining regional income. Income itself is a complex concept, and when referred to a region which usually has large extra-regional transactions becomes much more complex. As a result there are various ways of defining regional income depending upon the purpose of a given study. Perhaps, then, it is more pertinent to seek component data which can be more readily collected, and which subsequently can be aggregated into a regional income measure appropriate for a given study.

General discussion followed. Recognition was given to the problems of definition in data collection. At the same time, it was felt that the problem of definition was not serious in collecting certain kinds of data. For example, consumption expenditures, by type of commodity, by income and industrial classes, etc. for various regions could be obtainable (and it was strongly felt that they should be) without serious conceptual difficulties.

Data on savings and investment by regions was thought to be highly desirable, even though these raise problems of definition.

Discussion of other data, particularly data on regional production practices brought up the question of how general and widespread should the use of any set of data be to warrant the effort and expense at collection.

Capron:

Felt that production practice data would be very valuable, especially in regional development studies.

Chairman:

Seems to be general agreement that consumption data, production practice data and all kinds of component data for investment, savings and income would be valuable. Let us now turn to resource type data. Do current data on regional resources suffice for most purposes?

Ryan:

This would depend on what is meant by resources. Information available varies according to type of resource. Should certainly distinguish between natural resources and human resources

General discussion followed. Generally agreed that natural resource data e.g. mineral, waterpower and soil were satisfactory.

Marburg:

Pointed out that forest resource data were generally good, but could be collected in a way more suitable for use by other agencies.

Ryan:

Also indicated that it would be worthwhile to obtain better underground water data.

Discussion then turned to human resources. Generally agreed that this type of data was extremely important but generally inadequate. The extension of Social Security coverage, it was stated, is likely to yield more precise information on the labor force.

Establishment of contact with the Labor Market Analysis Committee of the Social Science Research Council was urged in order to coordinate efforts. It was also pointed out that a few persons already engaged in labor market studies are associated with us in regional research.

Buttrick:

Brought up the question of the adequacy of flow data and import-export data for areas. Should such data be collected on a county basis?

General discussion followed. Agreement that it was more advisable to correct existing imbalance in flow data rather than acquire data by finer areas (as counties). The I.CC. data on state-to-state shipments for individual and groups of commodities should be supplemented by similar data for water transportation, and certainly by some form of sample data on highway movements. Effort in this direction was judged to be more fruitful, too, because much data on a county basis could not be revealed because of disclosure regulations. On the other hand, it was also mentioned that flow data on the 1950 Census "economic area" basis could probably yield a finer breakdown without running into the problem of disclosure.

Hagen:

Raised what he considered to be an academic point. It would be extremely valuable to have continuous wealth data by regions in order better to interpret capital flows.

Capron:

Suggested such might be obtained in time from regional investment data, certainly to some extent from data on construction.

Wright:

Mentioned the availability of Dodge contract date on construction but seriously questioned its usefulness.

Chairman:

General agreement that in addition to better regional (county) income, survey, investment, consumption and production data, we should urge the collection of more data, or improved and more usable classifications of existing data, on regional (county) labor force, on a few particular types of natural resources, and on state to-state flow data of all kinds. Also, we should perhaps expend some effort at starting the process of accumulating physical wealth data by regions.

Buttrick:

Suggested setting up a permanent clearing house on regional data:

1. to furnish information on types of regional data currently available.
2. to gather and make known data requirements for regional research.

Palmer:

Concurred with Buttrick. Stated that the Bureaus of Business and Economic Research have similar clearing house problems. Suggested that we explore with their Association the possibility of working together on the clearing house problem.

General discussion followed. Liaison with government agencies such as the Statistical Standards Bureau and the Council of Economic Advisers was urged. Problems of setting up such a clearing house was discussed, especially the one of financing. It was generally agreed that a clearing house would be very desirable if it could be financed; and if the Social Science Research Council undertakes to support a Committee on Regional Economic Studies, this matter might be further discussed by such a committee.

Marburg and Hochwald:

Again raised the critical problem of priorities in data collection and classification, and in areas of research. Both priorities are interrelated. Hence, the need for further development and discussion of conceptual framework.

Chairman:

We might end our discussion on this note. One of our next informal sessions should, and I hope will, be devoted to the exploration of conceptual framework. After such exploration we ought to be in a position to indicate more concretely priorities in data collection and processing. The good news was that at the Midwest Economic Association meeting it was reported that the Social Science Research Council had reaffirmed its interest in regional economic research.

* * *

As an outcome of the data discussion at the April 20 informal meetings, there was a spurt of energy aimed to establish at the Federal Reserve Bank of Chicago a clearing house for regional information. After much communication and discussion with the officials of the Bank, the Bank's turndown of this request was most discouraging. Nonetheless, regional researchers continued to emphasize the need to have available an extensive empirical framework for testing hypotheses and provoking new ones. This concern of regional researchers, which has been persistent throughout the years, is present in the *Memorandum on a Census Monograph on the Location of Economic Activity and Its Relation to Population.* This memorandum is presented, as already noted, in Appendix A.

In all this work, W. Isard had positive support from a number of individuals: Joseph L. Fisher, Philip Neff, Edward Ullman, Everett E. Hagen, T.E. Grether, Harold F. Williamson, Stefan H. Robock, William G. Friedrich, to name a few—each of whom had primary obligations elsewhere and also lacked sufficient resources to mount a steady, continuing effort.

The next newsletter, dated May 1, 1951 indicates that after much effort a paper on "Regional and National Product Projections and their Interrelations" was added, albeit as the last paper, to a 1951 Conference on Research in Income and Wealth on Long Term Projections. At this conference scheduled for May 25–26, New York City, we were able to squeeze in another informal regional research meeting centered on Vining's path-breaking research outlined below, and a presentation by Leontief on emerging input-output research. Finally, the newsletter noted that it would be helpful if any scholar planning to attend the conference would indicate whether he/she has been invited to attend the conference. Professor Kuznets writes that he believes regional economists will be welcome at the conference, provided there are not too many of us.

The next newsletter was:

To Regional Researchers: May 31, 1951, Littauer 307, Harvard University. Cambridge, Mass.

Enclosed you will find an outline of the paper Professor Rutledge Vining presented the afternoon of May 26 at the informal meeting of regional economists attending the Conference on Research in Income and Wealth, New York. Under separate cover you have or will receive a copy of the Conference paper, *Regional and National Product Projections and Their Interrelations;* on pp. 41–47 is a synopsis of the paper Professor Wassily Leontief presented to regional economists the evening of May 25. A list of persons attending these sessions is appended.

It was my hope to be able to summarize in this letter the discussions following these excellent papers. However, this does not seem to be warranted since a good part of the discussions concerned points which merely required further elaboration by the speakers. This will be kept in mind in planning future sessions.

Spatial Aspects of the Structure and Functioning of an Economic System (outline of paper presented by Professor Rutledge Vining, May 25, 1951)

I. The Economy as a Flow System

1. Families and firms as economic units. Individual inflows and outflows and the continual management thereof.
2. Random changes in individual flows and in the system of individual flows.
3. Birth and death process of individual units and growth aspects of the system of units.
4. Fluctuations or evolutionary development of the flows within a system that are neither random or growth.

5. Flow interconnections between the units and the bearing of this upon the spatial arrangement of units.

II. The Spatial Structure of the System

1. Spatial dispersion of economic units not random. Intuitive conceptions of the pattern discernible in this arrangement.
2. Specification of the attributes of pattern in the study of systems of central places or cities.
3. Christaller's procedure in the study of systems of central places.
4. Development of a system of central places as a process with chance elements
5. A graded ordering of central places and systems of central places and empirical bases for classification.
6. Types of systems of central places and the geographic ebb and flow of money funds.

III. The Spatial Aspects of Economic Flows

1. A central place as a point of dispersion and point of absorption of economic flows.
2. Characteristic density pattern of destinations of traffic originating at a given point.
3. Characteristic density pattern of originations of traffic terminating at a given point.
4. Empirical data indicating the form and degree of stability of these relations.
5. Development of the parameters for these relations as aggregate flows expand and contract.
6. The conception of Regional Structure regarded as a pattern of density configurations relating to population and traffic and financial flows.

The newsletter of May 31 also listed (1) an upcoming roundtable discussion of the problems of regional research to be held at the Western Economics Association meeting, May 21–23, 1951 in San Francisco and (2) an informal session with regional sociologists and others at the American Sociological Society's annual meeting at Chicago, September 5–7, 1951. The latter was an outgrowth of the interest of regional economists expressed at the 1950 Christmas meeting to explore the possibility for a multidisciplinary approach to regional research, and particularly for embracing key sociological factors affecting regional development. Along with the June 11 and August 13 newsletters, invitations to attend were extended to Regional and Urban Sociologists, Human Ecologists and other interested persons including such leading sociologists as Howard Odum, Rupert Vance, Amos Hawley, Robert Angell and Dorothy Thomas.

The meeting proved fruitful, generating much interaction among social scientists. The condensed proceedings of it follows.

Condensed Proceedings of the Interdisciplinary Regional Research Meeting Held at Hotel Sheraton, Chicago, on Thursday, September 6, 1951

The participants were:

- Stuart A. Queen, Department of Sociology, Washington University, St. Louis, Mo.
- Carl F. Kraenzel, Department of Sociology, Montana State College, Bozeman, Montana
- Harry E. Moore, Department of Sociology, University of Texas, Austin, Texas
- Odin W. Anderson, Faculty of Medicine, University of Western Ontario, London, Ontario
- Walter Firey, Department of Sociology, University of Texas, Austin, Texas
- T. C. McCornick, Department of Sociology, University of Wisconsin, Madison, Wisconsin
- Donald Bogue, Scripps Foundation, Miami University, Oxford, Ohio
- Amos H. Hawley, Department of Sociology, University of Michigan, Ann Arbor, Michigan
- Gordon W. Blackwell, Institute for Research in Social Science, University of North Carolina, Chapel Hill, North Carolina
- Robert C. Stone, Urban Life Research Institute, Tulane University, New Orleans, Louisiana
- Alvin L. Bertrand, Department of Sociology, Louisiana State University, Baton Rouge, Louisiana
- Jerome K. Myers, Department of Sociology, Yale University, New Haven, Connecticut
- John P. Johanson, Department of Agricultural Economics, University of Nebraska, Lincoln, Nebraska
- Raymond E. Murphy, School of Geography, Clark University, Worcester, Massachusetts

- Robert S. Platt, Department of Geography, University of Chicago, Chicago, Illinois
- Harold M. Mayer, Department of Geography, University of Chicago, Chicago, Illinois
- Stuart Dodd, Department of Sociology, University of Washington, Seattle, Washington
- Howard G. Blunsman, Population and Housing Division, Bureau of the Census, Washington, D.C.
- H. H. McCarthy, Department of Geography, State University of Iowa, Iowa City, Iowa
- Henry D. Sheldon, Population and Housing Division, Bureau of Census, Washington, D.C.
- F. Stuart Chapin, Department of Sociology, University of Minnesota, Minneapolis, Minnesota
- N.J. Demerath, Institute for Research in Social Science, University of North Carolina, Chapel Hill, North Carolina
- And a number of others whose names were not obtained.

Chairman:

Opened the meeting with a brief statement on the work and activities of regional economists.

Purpose of meeting: to explore

1. whether there is an area for interdisciplinary activity in regional research;
2. if so, how cooperation between the several social sciences can be achieved in regional studies; and
3. for what specific purposes joint efforts can best be put forth.

Johanson:

Suggested that in order to get a common footing we might begin the meeting by having some of those present briefly discuss their current interest in regional analysis.

Others agreed and Johanson proceeded to discuss his problem of delineating regions in Nebraska according to patterns of gradients of farm population density, from the high density in the corn belt portion of Nebraska to the low density sand hills bordering the Great Plains.

Platt:

Long interested in problems of regions, primarily with respect to Latin America. Gradually has come to realize that agreement on regional boundaries is less important than using the regional device as a mode of approach. Regions of different sorts are required for different purposes.

Moore:

Spoke of the problem of school administration with which the University of Texas is now concerned. What kind of people become school administrators? How can the ability of an area to support education be measured? What is the best area for administration? These and other questions require the efforts of an interdisciplinary team.

Felt that such a team is necessary for any good area study. Thought we should formulate ideas to get people to work together. Pointed to problem of differing disciplinary languages.

Ullman:

Agreed with statement on language barrier. As a further point, was impressed by the separation of the contributions of each discipline, and the lack of knowledge of what was being done on a given topic by members of the several disciplines. Urged facilitating intercommunication.

Mayer:

Interested in focal or nodal regions, and urban areas as foci of interregional connections. Concerned with transportation which enables differentiation of economic activity in space.

Also interested in the problem of statistics for small areas. Felt that most data are for areas too large in extent, e.g. Census state economic regions. This restricts the usefulness of data. Efforts ought to be directed to getting data for smaller statistical units.

Bogue:

Economic areas were recently set up to establish small area data units. The county is much too small for many tabulations. The Census will group economic areas into sub-regions and regions. No other areal breakdowns are contemplated by the Census.

Queen:

Interested in Bogue's work. Also wants to get at the influence of metropolitan centers on surrounding areas, to study newspaper circulation, the attracting force of sport events, etc. In contrast to Bogue, and as a complement, is interested in extra-economic influences. Summed up by posing the question whether there are mutual influences between economic forces and extra-economic factors.

Kraenzel:

Interested in larger regional aspects. As a sociologist he has tried to describe dry land agriculture in the plains region—the machine aspect, the community and residential patterns, etc. Has found that adequate description and explanation requires all disciplines. Cited lack of adequate terminology.

Faucett:

The Federal Reserve Bank of Chicago feels that much more research on its region is needed. The Bank would like to have suggestions on what kinds of research it should pursue. Also noted the difficulty in learning about what others are doing.

Firey:

Interested in social organization and control as it pertains to spatial patterns of land use and regional resource management. Is seeking criteria for rational exploitation of resources. Cited, for example, agricultural practices which lead to precipitously falling water tables, which thus require sound regional conservation policy. Can a sociologist working with others in allied disciplines offer a generalized scheme to guide administration?

Chairman:

A number of us have mentioned the data problem in discussing our studies. Can we come up with any tentative recommendations on priorities in increasing the supply of data? Cited two possible ways to improve the effective supply of data:

1. collection of data which is obtainable but which is not being collected, and the correction of existing imbalances in data collection.
2. Development of methods of reclassifying and processing data already collected and of facilitating access to the data so as to make them more useful for regional analysis.

Kraenzel:

Because the plain states are parts of a number of census regions, the data are all torn apart, and are very difficult to use. Could such a situation be avoided? Further, resources to help collect and organize data are lacking in areas where no large cities exist. Could this imbalance be corrected?

Ullman:

As a partial consequence of the federal disclosure rule, data on urban phenomena, especially for transportation and industry, are much poorer than for agriculture and population, because the former activities are dominated by one or two large concerns in small areas. What can be done to alter this rule so that research does not suffer? Indications are that federal agencies are tightening up still more, and thus in effect withholding data on monopoly operations. This was not the intent of Congress.

One specific recommendation might be to have representatives of the public or pure research groups (as at universities) represented on various federal interdepartmental committees, such as on transportation statistics and metropolitan areas.

Mayer:

The census classification of industries is too minute for many purposes. This has the disadvantage of necessitating the use of the disclosure rule too frequently. Suggested reclassification to avoid invoking the disclosure rule so often. For example, he believed that "ladies hats made under contract" and "ladies hats not made under contract" are separate categories. For most purposes they do not need to be separate categories. But as a result, frequently it is not possible to get data on the broader group because of the disclosure rule on the smaller group.

Chairman:

Suggested that a recommendation be made to the effect that, where possible, grosser classifications of industries or subareas should be employed to reveal data at least for the larger group or area when the disclosure rule comes into effect on the smaller group or area.

Brunsman:

Disclosure rule helps Census obtain data. It fosters fuller cooperation by industry. Felt that suggestion of broader classification when disclosure rule operates should be given further consideration, and hoped that Census would be able to do so.

Rohrer:

Urged more consideration of cross-classification procedures in order to find a better and more meaningful set. Felt that much data collected by one discipline are under a heading that is not meaningful to another discipline, and hence is not effectively utilized. Also, because of differences in terminology, one discipline frequently overlooks valuable data collected by another discipline.

Kraenzel:

Commended the Census on its collection and processing of data by economic areas, but strongly urged that it not be at the expense of the county data.

A general discussion on the need for a clearing house and related facilities ensued.

Chairman:

We have now considered somewhat the data problem and have thrown out a number of suggestions for increasing the effective supply of data. Each of the suggestions should be carefully evaluated, and the whole problem thoroughly examined. This we cannot do now.

In this exploratory meeting perhaps it would be wise to turn at this point to conceptual problems in interdisciplinary research, to requisite analytical techniques, and to related matters.

Platt:

Felt that each regional problem would tend to be unique and different from other problems. Disciplines are not likely to get together wholesale, but rather on individual problems. In the light of this, is it possible to set up general procedures and techniques of analysis?

Kraenzel:

Some regional agency or research group with manpower and facilities should start off with some special project to create interdisciplinary interest.

Blackwell:

Felt that institutes already existing at several universities might well provide facilities and the spark for setting off such projects. For at such institutes, as at the University of North Carolina, there already exists a nucleus of interdisciplinary personnel cooperating with each other.

Bogue:

Agreed with Kraenzel and Blackwell. Wondered whether the Harvard project on the structure of the American economy might offer an opportunity for interdisciplinary cooperation. Asked the chairman to comment at some length.

Chairman:

Stated that there are a number of opportunities for interdisciplinary cooperation at the Harvard project and that probably there will be an increasing need for such as our economic analysis becomes further developed. For example, illustrated the need for the cooperation of geographers and sociologists.

Society is composed of many interdependent parts, whose interrelations can be approximately represented by a set of simultaneous equations. Thus far the project has been concerned primarily with the industrial complex of the United States society. We have been able to describe the interrelations of the outputs of various industries so that it is now possible to anticipate with considerable accuracy the effect of an increase in the demand for the output of any particular industry (say aircraft) upon the output of each of the other industries of the economy.

But a society is not composed of industries alone. It consists of regions, too. We have already experimented with what we call regional input-output analysis, and the results have been encouraging. Here we could benefit considerably from cooperative studies by geographers on the description and explanation of commodity flows from region to region. Also we need to know how the resources in the various regions shape the character of flows by imposing limits on outputs in any locality. And so forth.

But, also, society is not composed of industries and regions alone. It embraces a population, a population which is distributed among regions, a population whose character and tastes ultimately determine the outputs of all industries. A major channel for improving our results is through study of the consumption behavior of various elements of the population, classified by income, region, occupational status, ethnic group, place of residence (in terms of rural-urban, size of city, etc.), size of family, and so forth. Here, obviously, cooperative studies by sociologists are required for understanding better not only the economic and industrial development of various regions, but also, in turn, the complex of society itself.

General discussion followed in which the question of regional demarcation was taken up.

Platt:
Stated that there were two alternatives:

1. Delineate fixed regions
2. Collect data by small areas

Geographers have been studying the problem of the best set of fixed regions for years, but they have never reached any agreement. For this reason, felt we should confine ourselves to collecting data for small areas, and let the regional definition fit the particular problem and embrace the small area data accordingly. Most of the geographers in the country felt that the latter alternative was more desirable than the former.

Mayer:
Said that it was a step in the wrong direction to set up fixed regions of the nature that the Census uses and will use. Presented two reasons:

1. Regions should be delimited by the persons who are going to use the region in accordance with the use to be made of the regions.
2. With publication of data for fixed regions, there would be a tendency to cut down on publication for smaller units. This would frustrate studies for which a given set of fixed regions would be clearly inapplicable.

Brunsman:
Pointed out that it is not possible to give data on all sets of regions when disclosure rule operates, for example, on a county level. For if data were made available for all sets of regions, one might easily be able to derive data on a county level. Thus it is often necessary to use a set of fixed regions if any data of certain varieties are to be made available.

McCormick:

Indicated that in general sociologists would prefer having data for small areas and be free to choose from among the many possible sets of regions which could then be constructed.

Chapin:

Said that the situation is analogous to one which came up in a recent discussion on class structure. There it was pointed out that perhaps class structures were simply folk words and were not truly scientific entities. Employing folk words (and by implication, a set of fixed regions) would be analogous to an attempt to develop modern science with the materials labeled in folk terms only, such as the four elements, fire, air, water, and earth.

Moore:

Wished to shift the discussion somewhat to make two points.

1. Because of the work of Odum and others, the philosophy of regionalism has been well developed. The need for further effort here is not so intense.
2. Interdisciplinary cooperation ought to come about in terms of particular problems. A problem should be attacked by a given discipline. As the analysis requires cooperation of personnel from other disciplines, such should then be forthcoming. In short, complete interdisciplinary cooperation is not required in every problem.

Johansen:

Felt, however, that any regional analysis aiming at any degree of generality must be conducted on an interdisciplinary basis.

Kraenzel:

Made a final suggestion that we not overlook the point that it is important to take into account what people living in any given spatial unit consider their region to be.

* * *

The meeting was brought to a close by the chairman after he had asked whether or not it was the wish of those present that future meetings on interdisciplinary cooperation be arranged when feasible. It was the sense of the meeting that future sessions be arranged. The chairman was authorized to appoint a working committee on interdisciplinary cooperation in regional research to assist him in arranging such sessions and to provide guidance on whatever problems may arise on related activities.

Those interested but unable to attend (excluding individuals listed in previous memoranda)

- E.W. Burgess, Department of Sociology, University of Chicago, Chicago, Illinois
- Vincent H. Whitney, Department of Sociology, Brown University, Providence, Rhode Island
- Richard Meier, Program of Education and Research in Planning, Univ. of Chicago, Chicago, Ill
- Howard W. Odum, Institute for Research in Social Science, Univ. of North Carolina, Chapel Hill, NC
- Harvey S. Perloff, Program of Education and Research in Planning, Univ. of Chicago, Chicago, Ill.
- Julius Margolis, Survey Research Center, University of Michigan, Ann Arbor, Michigan
- Daniel O. Price, Institute for Research in Social Science, Univ. of North Carolina, Chapel Hill, NC
- Ernest M. Fisher, Institute of Urban Land Use, Columbia University, New York 27, New York
- Malcom Proudfoot, Department of Geography, Northwestern University, Evanston, Ill.
- A.J.Jaffe, Bureau of Applied Social Research, Columbia University, New York 27, New York
- Gerald Breese, Bureau of Urban Research, Princeton University, Princeton, New Jersey
- Walter H. Blucher, American Society of Planning Officials, 1313 East 60th Street, Chicago 37, Illinois
- Warren S. Thompson, Scripps Foundation, Miami University, Oxford, Ohio
- Marvin Sussman, Department of Sociology, Union College, Schenectady, New York
- Chauncy D. Harris, Department of Geography, University of Chicago, Chicago 37, Ill.
- J. Douglas Carroll, Community Service Center, 200 East Kearsley St., Flint, Michigan
- Rudolf Heberle, Department of Sociology, Michigan State University, East Lansing, Michigan
- Robin M. Williams, Department of Sociology, Cornell University, Ithaca, New York
- Louis Wirth, Department of Sociology, University of Chicago, Chicago, Illinois

- John Alexander, Department of Geography, University of Wisconsin, Madison, Wisconsin
- Preston James, Department of Geography, Syracuse University, Syracuse, New York
- Kingsley Davis, Department of Sociology, Columbia University, New York, New York

Returning to the newsletter of March 31, the most significant item was the result of the prodigious effort we mounted to obtain recognition by the American Economics Association of the importance of regional research. It announced that we had been able to schedule at the Christmas 1951 meetings a joint session with the American Economics Association (also jointly with the Econometrics Society) on regional and interregional analysis. This item testified to the fact that at least some economists had come to recognize the legitimacy of regional economics.

2.3 Era of Conceptual Thinking and Model Development with Multidisciplinary Explorations

After the May 1951 informal meeting in connection with the Conference on Research in Income and Wealth on Long-Term Projections of the National Bureau of Economic Research there developed another new major thrust of the group of regional researchers—an increasing interest in becoming acquainted with diverse conceptual frameworks being employed by regional researchers. The newsletter of July 25, 1951 states:

As you may recall, it was clearly evident at our May meeting in New York that regional economists need to become much more acquainted with each other's conceptual frameworks. The forthcoming Christmas meetings offer us another opportunity to familiarise ourselves with these frameworks. I think it would be very desirable to set up one or two (or even more) informal sessions devoted to the discussion of them.

To do this, however, requires the cooperation of each individual. In particular, I should like to know whether you have developed a conceptual framework for *general* regional analysis (to be distinguished from those which can be applied to only one specific region). If you have, would you be ready to distribute by November 15th to regional researchers copies of a paper embodying your framework, so that we may be prepared to discuss it thoroughly at Christmas?

At the formal session, under the joint sponsorship of the American Economic Association and the Econometric Society, there will be a panel and floor discussion of a paper by Rutledge Vining presenting his conceptual framework for regional analysis.

The next newsletters, of November 1 and November 13, 1951 provide evidence of the greatly increasing activities of the group of regional researchers.

To Regional Researchers, November 1 , 1951. Littauer 307, Cambridge, Mass.

In accordance with the wishes of many of those at the interdisciplinary meeting last September, I shall try to list, commencing with this letter, events and activities of interest to regional social scientists as well as to regional economists.

Of interest to all is the interdisciplinary session on regional analysis to be held the morning of August 7, 1952 at the annual meeting of the Association of American Geographers in Washington, D.C. This session is being organized by Professor Preston James, currently president of the Association.

On Saturday, November 17th, at 10 a.m., at the Southern Economic Association convention, Knoxville, Tennessee, there will be a round table on regional economics chaired by Stefan H. Robock (T.V.A.). R. Vining (University of Virginia) and W. Isard (Harvard) will deliver papers entitled, "The Need for Regional Economic Analysis" and "Current Developments in Regional Analysis", respectively. The discussants will be B. Ratchford (Duke), W. Hochwald (Washington Univ. St. Louis), H. Craven (Program Staff, Department of Interior) and R. French (Tulane).

On Friday, November 16th, at 4:30 p.m., at the same convention and same hotel, there will be an informal session to discuss the problem of *university programs for training in regional analysis*.

On Thursday, December 27th, there will be a round table on regional economics, jointly sponsored by the American Economic Association and the Econometric Society. Calvin Hoover (Duke) will chair the session, R. Vining (Univ. of Virginia) will deliver a paper on the "Spatial Aspects of the Structure and Functioning of the Economic System" and E. M. Hoover (Council of Economic Advisers), S. Robock (T.V.A.), H. Schwartz (Federal Reserve Bank of San Francisco), N. Wollman (University of New Mexico) and E. Dunn (University of Florida) will be the discussants. As I have mentioned previously, I am also hoping to be able to arrange one or more informal sessions on theoretical frameworks for regional analysis.

The Social Science Research Council is still considering the proposal for a committee on regional economic studies. It has appropriated a thousand dollars for an exploratory meeting of Western regional economists next Spring to be organized by E. T. Grether (Univ. of California, Berkeley).

Sincerely yours, Walter Isard.

To Regional Researchers, November 13 , 1951. Littauer 307, Cambridge, Mass.

This letter will be concerned with activities of regional economists. I am sending it, however, to all regional social scientists in the hope that some non-economists may find it possible to participate in these activities, and to keep all informed of each other's doings.

As you may recall, it seemed desirable to arrange one or more *informal* sessions on theoretical frameworks for regional analysis (to be distinguished from those frameworks which can be applied to only one specific region) at the forthcoming Christmas meetings of the American Economic Association and Allied Social Science Associations at Boston. In my letter of July 25th I asked to be informed as to whether or not you are developing any such framework. From the response to this letter and from subsequent inquiries, I gather that, aside from work on general theories of location, space-economy and optimum transportation systems, active research is currently being pursued on only two types of general frameworks: one, the spatial aspects of the structure and function of an economic system; the other, interregional and regional input-output analysis. Since we have been able to set up a formal round table on the former under the joint sponsorship of the American Economic Association and the Econometric Society (details noted in the previous newsletter), only one informal session has been arranged. This session scheduled for 4:30p.m., Thursday, December 27, 1951, Statler Hotel, Boston will consist of a short paper on *Current Developments in Interregional and Regional Input-Output Analysis,* after which I hope we can have a vigorous discussion of the virtues and limitations of this approach. It is urged that persons desiring to participate read in preparation the article: "Interregional and Regional Input-Output Analysis: A Model of a Space-Economy," *Review of Economics and Statistics,* November 1951. Additional background material on input-output analysis can be obtained from Wassily Leontief, *The Structure of American Economy,* Oxford Univ. Press, 1951.

Among other papers which may be of interest are, in chronological order:

1. Theories of Location and Trade, International and Interregional, by W. Isard and M.J. Peck, 2:30p.m., Wed., Dec. 26, Statler Hotel.
2. International Comparisons of Real Per Capita Income by J.B.D. Derkson, 9:30 a.m., Thurs., Dec. 27, Copley-Plaza Hotel
3. Interstate Migration in New England by N.L. Whetten and R.G. Burnight, 1:20p.m., Thurs., Dec. 27, Copley-Plaza Hotel

4. Criteria for the Establishment of an Optimum Transportation System by R.L. Dewey, and the Optimum Geographic Size of a Rate-making Unit for Electricity, by W.F. Kennedy, 2:30p.m., Dec 27, Statler Hotel
5. Some Considerations in the Use of the Residual Method of Estimating Net Migration, by J.S. Siegel and C.H. Hamilton, 9:30a.m., Fri., Dec. 28, Copley-Plaza
6. Regional Organization in the Sphere of Trade and Payments—Western Europe and the Sterling System, by J.M. Fleming, 9:30a.m., Sat., Dec. 29, Statler Hotel
7. Some Problems of Growth in Underdeveloped Countries, by A.I. Bloomfield, 2:30 p.m., Sat., Dec. 29, Statler Hotel

Sincerely yours, Walter Isard

Consistent with the objectives stated in the newsletter of July 25, there is presented in considerable detail the following proceedings. They do testify to another major advance in the scientific character of the group of regional researchers—namely, its aim to develop and dig deeply into conceptual frameworks and to be highly critical of diverse models that are advocated.

Proceedings of the Regional Economic Research Meeting at Statler, Boston, Thursday, December 27, 1951 at 4:30 P.M.

The participants were:

- G. E. McLaughlin, D.P.A., Washington 25, D.C.
- W. G. F. Friedrich, 601-19th St., Washington 6, D.C.
- H. U. Meyer, Göttingen, Germany
- Lt. Charles Hubbell, AFAPA-3., USAF, Washington 25, D.C.
- Robert L. Allen, Central Intelligence, Washington, D.C.
- R. S. Eckley, Fed. Res. Bank of Kansas City, Kansas City, Mo.
- G. H. Ellis, Fed. Res. Bank of Boston, Boston, Mass.
- C. M. Tiebout, Univ. of Michigan, Ann Arbor, Michigan
- Robert L. Steiner, 3443 Stettinius Ave., Cincinnati 8, Ohio
- William Gibelman, N.Y. State Dept. of Labor, New York, New York

- M. J. Peck, Harvard University, Cambridge, Mass.
- Roland N. McKean, RAND Corp., Santa Monica, Calif.
- Edgar Dunn, Jr., Univ. of Florida, Gainesville, Florida
- Harold G. Vatter, Champlain College, Plattsburg, New York
- E. Berman, Bureau of the Budget, Washington, D.C.
- R. P. Wolff, Univ. of Miami, Coral Gables, Fla.
- A.H. Rothstein, Univ. of Miami, Coral Gables, Fla.
- J. H. Craven, Dept. of Interior, Washington, D.C.
- W. H. Miernyk, Committee of New England, Boston, Mass.
- L. E. Fouraker, Penn. State College, State College, Pa.
- W. Baughn, Louisiana State Univ., Baton Rouge, La.
- F. Farnsworth, Colgate Univ., Hamilton, New York
- Leon Moses, T.V.A., Knoxville, Tenn.
- Martin Stoller, 1240 Park Ave., New York 28, New York
- A. T. Cutler, Fed. Res. Bank of Cleveland, Cleveland, Ohio
- R. Bowman, University of Pennsylvania, Philadelphia, Pa.
- A. Phillips, University of Pennsylvania, Philadelphia, Pa.
- M. T. Wermel, B.E.S., U.S. Dept. of Labor, Washington, D.C.
- A. A. Bright, Committee of New England, Boston, Mass.
- N. Wollman, Univ. of New Mexico, Albuquerque, New Mexico
- V. S. Whitbeck, Bank of New York, New York, New York
- R. French, Tulane University, New Orleans, La.
- C. Woodbury, Urban Redevelopment Study, Chicago, Ill.
- R. Grosse, Bureau of the Budget, Washington, D.C.
- Walther Hoffman, University of Münster, Münster, Germany
- V. Whitney, Brown University, Providence, R.I.
- T. Boyden, Harvard University, Cambridge, Mass.
- Walter Isard, Harvard University, Cambridge, Mass.
- Stefan Robock, T.V.A., Knoxville, Tenn. (Chairman)
- Robert Kavesh, Harvard Univ., Cambridge, Mass. (Secretary)
- and a number of others whose names were not obtained.

Chairman (Robock):

Presented a brief sketch of discussions at past meetings and introduced the speaker.

Speaker (Isard):

Explained that this meeting, in a way, tied in closely with the formal session of the morning. It seemed desirable at these meetings to discuss conceptual frameworks for regional analysis—and regional input-output technique is one such conceptual framework.

There followed a brief sketch of the model contained in the article "Interregional and Regional Input-Output Analysis, A Model of a Space Economy" in the *Review of Economics and Statistics*, November 1951.

The meeting was opened for discussion.

Grosse:

What evidence is there to say that interareal interindustry coefficients are stable?

Speaker:

Let's begin with the simplest case. Assume that there is only one source of a given ore and it lies in region I. Then, if an industry in region II using this ore expands by 10%, the flow of ore from region I to that industry in region II should increase approximately by 10%.

Now consider commodities at the other extreme which are purely local—e.g., power, various personal and business services, construction, brick, etc., where the service has to be performed on the spot or where the transport cost is so high that the commodity moves only short distances. It is clear that practically all requirements in any region of the outputs of these industries will be furnished by firms or persons in the same region. Here there are no interareal interindustry coefficients, but merely for each region a set of interindustry coefficients. This means that many cells in the table are empty and the question of the stability of interareal interindustry coefficients does not arise.

At this point the speaker was interrupted.

Berman and others:

Posed multi-source question of substitution. E.g., industry A in region I might supply industry B in region I, while industry A in region II supplied industry B in region II. Suppose that industry B in region I declined by 50%, while industry B in region II increased by 50%. Would not then industry A in region I supply (at least in part) the increased requirements of industry B in region II? This would mean unstable interareal interindustry coefficients.

Speaker:

If industry A in region II supplies industry B in region II because it has a definite transport cost advantage over industry A in region I, then there is no reason to expect that with increased output of industry B in region II that any of its inputs will come from industry A in region I—with the exception, of course, where industry A in region II initially may have been operating at full capacity and where the increase in capacity lags for one reason or another. Where the capacity factor is introduced, the same problem of substitution between 2 industries in supplying a 3rd one also arises in the national model.

Steiner:

Pointed out the fact that many firms in a particular area trade with other firms in the same or other areas, not so much with respect to transport cost savings, but with respect to reliability of supply and service. This, then, would tend to detract from the usefulness of constant coefficients.

Speaker:

To the extent that firms have fixed ties with other firms—institutional or otherwise—i.e., continue to purchase from the same group of firms, then relations are stable. It is only when a firm in a given area shifts its purchases back and forth from one firm in one area to another firm in a different area that the coefficients become unstable.

Then ensued considerable discussion about the stability of the interareal interindustry coefficients. It was brought out that the limitations of the assumption were greatest in the case of goods which are produced and consumed at many places and which, because they have high value and incur small transport cost, tend to be shipped all over the nation.

Vatter:

Posed the problem of regional development in upper New York State. Stated that a static, two-dimensional approach was not helpful at all in economic development problems such as anticipating the effect of working new titanium deposits or bringing new industries into an area.

Speaker:

Presented additional material on the model as contained in the article cited above. Stressed the importance of the Bill of Goods sector, not only to embrace the unexplained sectors of the economy, but also to include those sectors in which direct interregional shifts of industry are to be expected. The working of new titanium deposits would be such a case of an interregional location shift. The development of the Trenton iron and steel works would be another. In other words, in the Bill of Goods sector, a good part of the locational shifts associated with major industries must be inserted.

Considerable discussion ensued again as to how much was to be put in the Bill of Goods sector and how much was to be left in the structural matrix.

Boyden:

Thought we had underestimated the usefulness of the table. Though analysis was required to set up the Bill of Goods for any economic development process, the requirements of certain industries such as power might be indicated by the structural matrix. Thus, even for dynamic situations the framework is useful when cautiously employed.

Ellis:

Raised question of whether technological change could be handled by the model.

Speaker:

As the model exists now the answer is no, but models are being developed on a national scale which are attempting to attack the problem of technological change as well as other dynamic aspects.

Robock:

Wondered if Ellis' question was relevant. No one has yet been able to handle problems of technological change. Thought the fact that the regional input-output model could not deal adequately with technological change was not a major argument for not using it.

Speaker:

Explained that, if technological changes were known beforehand, it was possible to alter the coefficients to account for such changes. Urged caution and moderation in anticipating what the model will be able to do. First, results cannot be any better than the data available. Second, the technique cannot substitute in any way whatever for the usual location and market analysis, which is required to set up adequately the regional breakdown of the Bill of Goods. Also felt that the best use of input-output was for anticipating repercussions of small changes only.

Wollman:

Stated that in many cases small changes could involve greater instabilities in coefficients than moderate changes. The Texas steel industry might get its coal from Oklahoma and New Mexico—however, the additional coal for a small increase in the industry might come from Oklahoma alone.

Speaker:

Admitted this possibility, but questioned to what extent such a situation might generally occur. Thought that we really couldn't gauge the extent of such instability until all the required data had been assembled.

Allen:

Returned to the problem of constant coefficients. Suggested that a function relating the input of any industry in any region to another industry in a different region might be used instead of an interareal interindustry coefficient. This would involve considerably more work if it could be done.

Asked the speaker whether regional input-output techniques might be better for planned rather than competitive economies—also whether interareal interindustry coefficients can be deduced theoretically from the knowledge of the industrial structure of each region above.

Speaker:

Felt that he could not answer the question on competitive versus planned economies. Seriously doubted whether reliable data could be deduced. Again stressed the point that results cannot be better than data which enter into the model.

Further discussion followed.

Robock:

Regardless of whether or not the coefficients are constant there is real use in the structural matrix as it stands. Felt that for T.V.A. problems the consistency framework which the matrix provides and the static picture of the interrelation of industry were valuable. Was not really concerned with the problem of and limitations to mechanical projection.

Berman:

Shifted to the problem of computation. Thought that the disadvantage of increased computation involved in setting up a regional input-output model would be quite serious.

Moses:

Answered Berman and said that for a given number of industries regional input-output analysis requires increased computations. However, it must be remembered that spatial disaggregation permits greater accuracy of results than no disaggregation of the nation at all. Therefore *spatial* disaggregation is in part a substitute for industrial disaggregation. It is likely that just as much accuracy can be obtained for the same amount of computational effort by considering fewer industrial breakdowns under conditions of spatial disaggregation than a finer industrial breakdown with no spatial disaggregation.

Wollman:

Brought up one final point before closing: How are regions defined?

Speaker:

There is considerable flexibility in selecting regions to be considered. To illustrate this point he referred to the setup visualized for T.V.A. study. 201 counties in the power distribution area can be set aside as one region, the remainder of the 7 Southern states in which the 201 counties lie as a second region, and the rest of the U.S. may be split up into regions in any way which would be most useful for T.V.A. purposes. For another type of regional study an entirely different breakdown of the U.S. might be employed. Again, any demarcation of regions must be considered with the desired industrial breakdown and the limitations of the data in mind.

Chairman:

In bringing the meeting to a close asked for suggestions for subjects to be discussed at future meetings. Three were suggested:

1. Reexamination of methodology of various approaches to regional analysis.
2. The need for regional research; i.e., the various types of research projects which are highly desirable and should have priority on personnel and financial resources.
3. Discussion of the possible applications of interregional and regional input-output analysis from a positive (rather than from the above negative) standpoint, and an evaluation of its usefulness for problems peculiar to states and local areas.

* * *

Despite the push for scientific advance involved in developing and critically evaluating conceptual frameworks and models, the group of regional researchers continued to desire to explore intensively multidisciplinary approaches. This is clearly evident in the selected materials of the largely self-explanatory newsletters of January 25, April 11, July and August 19, 1952. They announced:

1. An interdisciplinary meeting on metropolitan regional research to be held at Baltimore, on April 25 at 2 p.m. at the convention of the American Institute of Planners. The participants in the discussion will be: Paul Oppermann (planner), San Francisco; N. J. Demerath (sociology), University of North Carolina; Victor Roterus (geography), Department of Commerce; W. Isard (economics); with Robert B. Mitchell (planner), University of Pennsylvania as chairman of the session.
2. The Western exploratory meetings on Regional Economics which have been organized by Dean E. T. Grether (University of California) and which are not planned as a full scale conference, but rather as meetings of committee size to take place May 21–23, Berkeley, California.

   The titles and authors of the papers presented were "Capital Formation in the Mountain West" by Morris E. Garnsey; "Economic Theory and Regional Analysis," by J.A. Guthrie; "Some Emerging Concepts and Techniques for Regional analysis," by W. Isard; "Identification and Measurement of an Industrial Area's Export Employment in Manufacturing," by Philip Neff and Robert Williams; "Competitive and Complementary Economic Development Among the Eleven Western States" by Elroy Nelson; "Estimating Regional Balance of Payments in the Pacific Northwest" by Paul B. Simpson and Shirley Burr; and "The 'Spark' Starting Regional Development with Special Consideration to a Hypothesis on the New Role of Amenities" by Edward L. Ullman.

3. An *Area Studies Panel Discussion* in part based on an interdisciplinary seminar led by Preston E. James in March at Harvard planned for the annual meeting of the Association of American Geographers scheduled for 9:30 a.m. Thursday, August 7, Hotel Statler, Washington, D.C. The specific contribution of various social science disciplines to a program of area research will be discussed by: Edward A. Kennard, *Anthropologist* (Foreign Service Institute, Department of State); Howard W. Odum, *Sociologist*, University of North Carolina; Edward L. Ullman, *Geographer*, University of Washington; and W. Isard, *Economist.* Preston E. James will preside.

   Since a number of regional scientists find informal meetings more fruitful than programmed discussions, a bull session on *Theory and Analysis in Regional Studies* is being scheduled for the afternoon of August 7 beginning at 4:30 p.m. in Council Room, Hotel Statler, Washington, D.C.

4. Another informal meeting to discuss *Ecology and Regionalism* to be held on Thursday, September 4, 1952, in Atlantic City, New Jersey, at 4:30 p.m. in connection with the annual convention of the American Sociological Society (Sept. 3–5).

A formal interdisciplinary session on *Regional Studies* is also scheduled by the American Sociological Society and the Population Association of America for 1:30 p.m. Friday, September 5. Rupert B. Vance (University of North Carolina) will chair the session which will consist of the following papers: "Social Change in the Southern Appalachian Region" (by Paul F. Cressey, Wheaton College), "World Regions and the Correlates of Urbanism" (Kingsley Davis, Columbia University), and "Economic Processes in an Urban Industrial Region" (W. Isard, Harvard). The discussants will be: Howard W. Beers (University of Kentucky), Albert J. Reiss (University of Chicago), and Joseph Fisher (Council of Economic Advisers).

Testimony to the advanced thinking of leading geographers (at the Association of American Geographers meeting) with regard to basic multidisciplinary regional research and more specifically the role of geographic theory is in the following notes.

Notes on the Regional Research Meeting
Held at Hotel Statler, Washington, D.C. August 7, 1952 at 4:30 P.M.

The participants were:

- Edward L. Ullman, University of Washington, Seattle, Washington
- Allen Rodgers, Penn State College, Penn State, Pa.
- E. F. Penrose, Johns Hopkins University, Baltimore, Md.
- Raymond Murphy, Clark University, Worcester, Mass.
- Chauncy Harris, University of Chicago, Chicago, Ill.
- Thomas R. Smith, University of Kansas, Lawrence, Kansas
- H. H. McCarthy, University of Iowa, Iowa City, Iowa
- Eurgene Orr, Department of Agriculture, Washington, D.C.
- Harold Mayer, University of Chicago, Chicago, Ill.
- Alexander Melamid, New School for Social Research, N.Y.C.
- W. G. Friedrich, Washington, D.C.
- Rupert Vance, University of North Carolina, Chapel Hill, N.C.
- Theordore Shadbad, RAND, Santa Monica, Calif.
- Gordon Reckord, Mutual Security Agency, Washington, D.C.
- Frank Hanna, Duke University, Durham, N.C.
- Robert Platt, University of Chicago, Chicago, Ill.
- Peter Nash, Boston Housing Authority, Boston, Mass.
- Dorothy Muncy, Washington, D.C.
- Victor Roterus, Department of Commerce, Washington, D.C.
- Gustav Larsen, Department of Commerce, Washington, D.C.
- Walter Isard, Harvard University, Cambridge, Mass.
- Howard Green, Stop and Shop, Boston, Mass., *Secretary*
- And several others whose names were not obtained.

The discussion was started off by *Ullman* who pointed out that all geographers are basically interested in regional research. Recognizing that geographic studies of regions are for the most part descriptive, he led to the fundamental question to be discussed: *is there a place for regional theory in geography?*

Roterus:

Agreed that theory in geography was lacking, but suggested that perhaps a unified social science theory of regional structure was a more basic requirement than a theoretical framework for geography.

It was indicated that there is some integration and coordination of social scientists in the field of urban studies where urban geographers, urban land economists, city planners, urban sociologists and other students of cities work together without regard for the limits of their own disciplines. It was suggested that perhaps application of theory could proceed at a faster rate where it was integrated.

Mayer:

Posed a basic obstacle to the development of geographic theory. To a geographer every region is unique. The very fact that each region occupies a unique spatial position differentiates regions from one another. If regions then are unique, what is the basis for theoretical generalizations?

The ensuing discussion raised several interesting points. Perhaps geographers were approaching regional study with a wrong point of view. Geographers have always looked for differences. Perhaps for a while they should look for similarities, and, after having developed a conceptual or theoretical framework for handling similarities, return to the study of regional differences.

Vance:

Suggested that an area of agreement among regional studies in the various disciplines might be the ecological approach.

Melamid:

Then advanced the belief that geographers should start from the consideration of pure space *per se* (a common characteristic of all regions) and then introduce the various factors which tend to make regions dissimilar. This theoretical start (as we have from Thünen, Lösch and Waibel) could then become a bridge for the study of human ecology.

Isard:

Raised the question of how the work of Stewart, Zipf and Reilly, which treats pure space through considering only given populations and the distance variable, might be used in the theoretical approach to nodal (or functional) regions.

Out of the discussion of Stewart's $\frac{P_1 P_2}{d}$ formula which expresses interaction between two populations at a distance, there emerged a general feeling that the interaction process was basic to the understanding of intra- and interregional relations. But why should the simple Stewart type formula be the point of departure?

Rodgers:

Cautioned that there should not be too much reliance on the theoretical approach. For example, the theoretical economic approach to location cannot evaluate the personal factor in the location of industry, which factor in many cases dominates a decision.

Friedrich:

Felt the study of regions as a set of fields of forces, following some of the concepts of the physical sciences and along the lines of Stewart and Zipf, is a very useful general approach and avoids the perplexing problem of defining regional boundaries.

The question was raised; is it not possible to begin with a purely spatial approach, such as is implied in the work of Stewart and Lösch and which involves oversimplified theoretical constructs which abstract from much of reality, and then introduce in more and more refined fashion the differentiating elements of regions? If one were to begin with Stewart's hypotheses (which might be appropriate for nodal regions, but not for homogeneous regions), could an aggregative variable measuring the effect upon interaction of a region's total resources be introduced to qualify the effect which stems from the distance variable alone? And with further study, especially through contrasting "computed" results with actuality, could not the aggregative variable be decomposed into several variables which would more adequately describe the complex differentiations among regions and their resulting interactions.

Platt:

Suggested that we are attempting too much at one time by trying to fit regions into a model. Regions are subjective and too complex for a common mould. Regions would vary according to the criteria used. Criteria therefore becomes the limiting factor.

(Platt's remarks started off a lively discussion on the concept of the region and its purpose. Since many of the points made have already been mentioned in the notes of previous meetings, they are not recorded here.)

The question was raised whether geographers or a group of social scientists working together should be able to predict, to some extent at least, the amount, location and pattern of land uses that would result if an uninhabited area were to be settled by human population. If such prediction were possible, it was felt that it would have to be couched in terms of a set of possible types of development, the type that might emerge being dependent upon the political system and cultural values associated with the population.

Penrose:

Felt that the conceptual framework of Ohlin was particularly valuable in this connection, that physical factors such as soils and climate could be meaningfully classified and studied for their economic effects.

Could the effect of a particular type of soil be objectively evaluated in order that the soil variable might be introduced into a theoretical framework to quantify interaction of regions via the distance variable alone (as suggested above)?

Platt:

Remarked that even soils and climate are subjective factors and not easily measurable. Their significance in any region depends upon the cultural institutions and the state of technology. And as Ullman pointed out, geography by its nature is ever-changing.

Vance:

However felt that to some extent, in a particular culture and at a given point of time, subjective and objective factors can be evaluated together.

The discussion ended with no agreement among participants as to the role of theory in regional studies and in geography. Nonetheless, as is clearly evidenced by the notes, the discussion cleared up some of the basic issues which need to be faced by those who advocate and embark upon the construction of theoretical models for regional study.

* * *

Enclosed as Appendix C is a memorandum by Lee Benson (Department of History, Columbia University), which should be of interest to many of us. For those of us who are concerned with analysis of regional development in the United States its discussion of historical regional data and of specific historical regional studies are of obvious relevance. For others who are concerned with development of various regions of the world, particularly underdeveloped ones, this memorandum should be of equal interest. Though there are dangers to comparative analysis if pushed too far, and though even seemingly small cultural differences may preclude direct transfer of results from one research project to the next, nonetheless study of the historical development of the regions of the United States, of the causes of their differential rates of growth, of the manner in which the capital obstacle was overcome and the transport net established, etc., can yield penetrating insights into the process of development and desirable policies re: the economic and industrial growth of many less developed regions of the world. Walter Isard.

With regard to the several regional sessions scheduled at the 1952 Christmas meetings, the newsletters of October 22 and December 12 provided the following information.

A session on *Regional and Spatial Economics* sponsored by the Econometric Society is scheduled for December 27, 9:30–12:00 A.M., Conrad Hilton Hotel, Chicago. The papers are:

1. A Spatial-Equilibrium Model of the Livestock-Feed Economy in the United States, by Karl Fox.
2. The Equilibrium of Land-Use Patterns in Agriculture, by Edgar S. Dunn.
3. Regional Economics and Economic Development, speaker to be designated.
4. Discussants will be Guy Freutel and Gregory Vore. W. Isard will chair.

A session on *Interregional Analysis and Regional Development* sponsored by the American Economic Association is scheduled for December 29, 9:30 A.M., Hotel Conrad Hilton, Chicago. The papers are: "Interregional Commodity Flows," by W. Isard, and "Fuel, Power and Regional Industrial Development in Canada" by John Dales. Penelope Hartland Thuberg and David Schwartzman will be discussants; chairman to be designated.

In order not to conflict with the above sessions, and also to allow time for travel from Chicago to St. Louis, a third session is scheduled for 10 A.M., December 30, Jefferson Hotel, St. Louis in connection with the meetings of the American Association for the Advancement of Science and jointly sponsored by Sections E, H, K and O. The session is entitled: *Regional Research: Emerging Concepts and Techniques* with papers representing the fields of:

1. Geography, by Edward A. Ackerman.
2. Political Science, by Ernest A. Engelbert.
3. Planning, by Harvey S. Perloff.

Edgar M. Hoover and Byron T. Shaw will be discussants.

As a result of the stimulating discussion on theory in geography and regional studies at the informal session of regional researchers held last August, there will be scheduled a formal session on the same subject at the forthcoming convention of the American Association of Geographers, March 30-April 2, Cleveland.

Altogether, the newsletters of 1952 clearly testify to the entrepreneurial activity of regional scholars and vigorous development and spread of interest in regional research and problems. City and regional planners, geographers, and demographers were effectively added to the sociologists and economists of 1951, and some ties with political scientists were effected. Moreover, contact with the American Association for the Advancement of Science became extensive. Yet it was also clear that the historian's approach as related to regionalism, presented in the excellent memorandum circulated with the newsletter of October 23, 1952 (see Appendix C), was not likely to be embraced in the core research of the developing group of scholars. The failure of historical analysis to provide models leaning toward both mathematical formulation *and* quantitative testing was not present at that time; however, the strength of such analysis is its avoidance of the employment of assumptions, explicit and implicit in the work of many non-historians—assumptions which tend to or do belie reality.

Note that the *informal* meetings in 1950 and 1951, at times when the American Economics Association held its annual convention, resulted in stimulating, constructive discussion and the identification of critical areas for creative research. They tended to be much more valuable than regular formal sessions programmed at the American Economic Association meetings. This was so since the President of that Association was in charge of that Association's program and thus exercised considerable influence on the emphasis and choice of participants in the regular regional research sessions scheduled. Thus, this situation led to the desire of regional researchers to consider seriously the possibility of forming an association, whereby they would have full control of the program at an annual convention

Also, note that there was not distributed in 1953 a memo on proceedings of an informal meeting of regional researchers at the December 1952 meetings of the American Economic Association. Probably, such a meeting did not occur. As reported in the newsletter of December 12, 1952, there were three formally sponsored regional meetings and thus little need for an informal one.

Moreover, by the end of 1952 the objective of getting together at informal meetings to become familiar with research aims and projects that each researcher was engaged in had been largely achieved. There then emerged a need to learn at considerable depth about the research procedures and approaches of some of the leading figures in regional research. This need was already reflected in (1) the scheduling of the paper by Vining at the May 1951 meeting of the Conference on Research in Income and Wealth, (2) the arranging for discussion of Isard's research at the 1951 meetings, and (3) the distribution of Benson's paper along with the October 23, 1952 newsletter and the papers by Ullman, Perloff, Ackerman, Engelbert and Robock and Ruttan along with, respectively, the December 12, 1952 and the February 23, April 23, June 9 and August 19, 1993 newsletters. All this testified to the fact that regional research had aspired to and already was achieving another more advanced level of scholarship.

The newsletters of 1953, as those of 1952, indicate the growing outreach of regional researchers. There were meetings and formal sessions with geographers, planners, sociologists, economists and econometricians. These were associated, respectively, with the annual conventions of the Association of American Geographers, the American Institute of Planners, the American Sociological Society, the American Economic Association, and the Econometric Society; and there was an informal session scheduled at the convention of the American Political Science Association.

Additionally, there began to develop, in the eyes of a number of regional scholars a multidisciplinary regional research area that had a unique character—one that might begin to qualify as a unique field of study. One aspect of this development was the use of mathematics and models well beyond what geography was aspiring to embody in its scope. Another was the extensive attention given to nodal (urban and metropolitan) regions and the gravity-type concepts which were very strange to economists (considered to be quackery by some) as well as the notion of distance inputs (later redesignated transport inputs). Still another was the extensive use of economic analysis in city and regional planning analysis, which involved diverse factors and interdependencies well beyond regionalism as conceived by sociologists such as Odum and Vance and human ecologists such as Hawley. One more was the use of regional and interregional input-output and other techniques to study population settlement and movement that were distinct from the approaches of urban sociologists, human ecologists and demographers. This pulling together of new ideas, concepts and frameworks percolating in several fields and integrating them to form more applicable findings and policy suggestions has been and continues to be the unique feature and strength of the field of regional science—a field that was to come to develop rapidly with multidisciplinary nuances and features—a field that became well defined by the publication of Methods of Regional Analysis (Isard, et al., 1960) and further developed in Methods of Interregional and Regional Analysis (Isard, et al., 1998).[9]

At this point, it is pertinent to provide a more specific account (and illustration) of the major impact upon urban and regional analyses of the emerging regional science. Take the impact upon city planning starting with the early 1950s. Then the problems associated with urban land use were extremely acute for city planners. Typically, they zoned land for diverse types of uses (industrial, commercial, and residential) according to subjective criteria such as "perceived" visual qualities of the several possible spatial arrangements; compatibility in terms of smells, noise, and accessibility needs; rules-of-thumb procedures to check on adequacy of tax base; and various other subjective factors. The only economics in use at that time

---

[9] It should be noted that geography dipped considerably into what became regional science under the energy and stimulus of William L. Garrison and his students in a branch designated quantitative geography. But this thrust lapsed when it failed to be given proper encouragement and support by those in the rest of the extensive field of geography, in part as a result of the growing involvement of geographers in GIS (geographic information systems).

was the standard Homer Hoyt basic-service ratio of one-to-one, widely employed in order to estimate the impact of new industry upon employment in service trade. There was no real look at interindustry interconnections, nor at relationships of industry with diverse commercial, residential, and cultural activities. Here was the area where the emerging regional science, via regional input-output analysis, had its first impact. For we were able to indicate to planners how one could go from an exogenous change in an industrial sector (such as the construction and operation of an integrated iron and steel works), with which they had to constantly deal, to a systematic projection of (1) effects upon the output and employment of each of a number of other important industrial sectors, (2) new income simultaneously generated, (3) subsequent new demands for retail and commercial services, and for housing by different income groups, and so on. Thus they were able to begin to construct more detailed, yet consistent, land use plans. At this point we were not yet concerned with an overall *consistent projection* of the future of a region or an urban metropolitan area. That came later. We first focused on *consistent impact analysis.*

Shortly thereafter, another critical problem of planners came to have an important impact on the development of methodology: the problem of transportation planning. If we set aside adequate land for new industrial developments, new housing, new commercial activity, and new cultural activities that might be associated with some basic new industry (such as an integrated iron and steel works), how could we project the amount of traffic that would be generated—journey-to-work phenomena, shopping trips, commodity flows, and so on?

The planners recognized that this was necessary in order to identify where new transport lines and capacity should be constructed to avoid congestion and breakdown of the transportation system. To answer this question we began to explore the potentials of the gravity model. The classic article was by Gerald A. P. Carrothers (1956); Doug Carroll (1955) and Alan Vorhees (1955) were the pioneers in its application. We were in fact the first social science to legitimize gravity and related spatial interaction methods. Distinguished economic theorists who used to scoff at these models now seriously employ them.

But it was soon evident that the use of a gravity model as a supplement to, or as a simple addition to, an input-output or other model was inadequate. We came to recognize that we needed to adopt a more effective "systems" approach in our work—that is, to start to make consistent projections of the magnitudes of the subsystems forming the metropolitan region. In less sophisticated terms, we soon recognized that if a transport network is changed to meet some congestion problem, or to fill in some gap in the network—and, of course, this was a main concern of planners and regional economists—it does not follow that everything else stays the same, as was usually assumed in the existing 1950 analyses although lip service was given to the fact that things do change. We thus reached the stage in the middle 1950s where the need to conduct systematic transport studies to meet the congestion problems of metropolitan areas, which had become foremost, required us to interconnect the transportation subsystem with the industrial location, the residential location, and the commercial location subsystems. It was recognized that a model of each subsystem had to feed important inputs into the other subsystems in order to project their magnitudes, and in turn required feedback information from them. Once again our contributions were to be primarily technique-oriented. We still had not caught up with the most advanced methodology employed in the other social sciences and particularly economics.

It is to be noted that the work of Benjamin H. Stevens on the subsystem interconnection problem was most significant. After theoretically developing regional and interregional linear programming with Isard he went on to write the classic article with John Herbert (1960), "A Model for the Distribution of Residential Activity in Urban Areas." That model set off a new wave of model building in regional science and formed the conceptual framework for most, if not all, of the major transportation studies of the '60s and '70s. Interregional linear programming came to the forefront and with it were fused the input-output, comparative cost, industrial complex, and gravity models that had already been developed and partly synthesized in applied research.

The above account of the specific impact of the emerging regional science upon city planning involved a succession of advances in techniques. The associated research activity came to embrace over time the increasing fusion of models. Such an outcome was replicated with regard to many other specific problem areas and subfields attacked by the emerging regional science.

Returning to the mid-1950s, perhaps the most significant public activity of regional researchers was their deep involvement in the organization and presentation of papers at the Conference of the Committee of New England of the National Planning Association and the New England Council which was widely reported upon by the media. Also, this conference was held in conjunction with the meeting of the AAAS (American Association for the Advancement of Science) on December 27-30, 1953 in Boston. (Part of the program is presented below.) There the existence of the regional research group and its potential for basic applied research was recognized by a number of national leaders and distinguished scholars of diverse background. No longer could the group be considered to be a splinter unit.[10]

[10] There is no record of an informal business meeting of regional researchers at the December 27–30 get-together. This might have been the result of their intensive interactions with several groups noted above.

Selected Sessions from the Joint Conference of the Committee of New England of the National Planning Association, New England Council, and the American Association for the Advancement of Science (AAAS), Boston, December 27-30, 1953

SUNDAY AFTERNOON: CONCURRENT SESSIONS 1 AND 2, DECEMBER 27

2:30 P.M. Concurrent Session 1
Symposium: The Scientists in American Society, Part I; Freedom for Scientific Inquiry.
Arranged by a Committee of Section K—Social and Economic Sciences, Conrad Taeuber, Secretary.

Detlev W. Bronk, Presiding

Papers: 1. The Beliefs and Expectations of the Public
Clyde W. Hart, *Herbert Hyman*, *Paul B. Sheatsley*, and *Shirley A. Star*, National Opinion Research Center, Chicago, Illinois.

2. The Social Psychology of Political Loyalty in Liberal and Totalitarian Societies
Raymond A. Bauer, Russian Research Center, Harvard University.

2:30 P.M. Concurrent Session 2;
Symposium: Regional Analysis.
Joint session of AAAS Section E—Geology and Geography and AAAS Section K—Social and Economic Sciences.

Walter Isard, Presiding

Papers: 1. The Economic Structure of the Metropolitan Region
Robert A. Kavesh, Darthmouth College

2. Regional Advantage in Aluminum Location
John L. Krutilla, Tennessee Valley Authority

3. Labor Costs as a Locational Factor in New England Industry
William Miernyk, Northeastern University.

MONDAY MORNING, DECEMBER 28

9:30 A.M. Subject: *Symposium: The Economic State of New England, Part I*
Joint program of AAAS Section K—Social and Economic Sciences, the Committee of New England of the National Planning Association, and the New England Council; co-sponsored by AAAS Sections E—Geology and Geography and M—

Engineering. Arranged by *Sumner H. Slichter* and *Walter Isard*, Harvard University;
A presentation of the Committee of New England.

Edwin D. Canham, Editor, Christian Science Monitor, Presiding

Papers:

1. Spotlight on New England
H. Christian Sonne, Chairman, National Planning Association Board of Trustees

2. Transition in New England
George H. Ellis, Director of Research, Committee of New England

3. New England's Natural Resources
John Chandler, New York and New England Apple Institute, Inc.

4. New England's Human Resources
Arthur A. Hauck, President, University of Maine

MONDAY AFTERNOON, DECEMBER 28

2:00 P.M. *Symposium: The Economic State of New England, Part II.*

Sumner H. Slichter, Presiding

Papers:

5. Transportation and New England
Robert M. Edgar, Vice President, Boston and Maine Railroad

6. New England's Financial Resources
Joseph A. Erickson, President, Federal Reserve Bank of Boston

7. Management and Research in New England
Earl P. Stevenson, President, Arthur D. Little, Inc.

8. Taxes, Public Expenditures, and New England
Alden C. Brett, Treasurer, Hood Rubber Company

MONDAY EVENING, BANQUET SESSION, DECEMBER 28

7:00 P.M. *Symposium: The Economic State of New England, Part III*

Laurence F. Whittemore, President, The New England Council, Presiding

Papers:

9. Goals for New England
Leonard Carmichael, Chairman, Committee of New England, and Secretary, Smithsonian Institution, Washington, D.C.

10. New England Viewed in Perspective
H. Christian Sonne, Chairman, National Planning Association Board of Trustees

TUESDAY MORNING, DECEMBER 29

9:30 A.M. Subject: *The Individual Scientist in Today's World*; arranged by Victor Paschkis, Heat and Mass Flow Analyzer Laboratory, Columbia University, Chairman of the Education Division of the Society for Social Responsibility in Science.

Victor Paschkis, Presiding

Papers:

1. The Scientists' Responsibility for Interpretation of Concepts to Laymen
Kirtley F. Mather, Professor of Biology, Harvard University
2. The Scientist as Architect of Social Change
Stuart Mudd, Professor of Microbiology, University of Pennsylvania

---

TUESDAY AFTERNOON, DECEMBER 29

2:00 P.M. *Symposium: Scientific Research and National Security*; Joint session of AAAS Section K—Social and Economic Sciences, AAAS Section M—Engineering. and the National Academy of Economics and Political Science, with the collaboration of the National Social Science Honor Society, Pi Gamma Mu. Arranged by Amos E. Taylor, Pan American Union.

Lowry Nelson, Univ. of Minnesota, Presiding

Papers:

1. The Role of Government in the Coordination of Basic Research
Alan T. Waterman, National Science Foundation
2. The Contribution of Industrial Research to National Security
Mervin J. Kelly, Bell Telephone Laboratories
3. Scientific Research and the National Economic Potential
J. Carlton Ward, JR., Vitro Corporation of America

TUESDAY EVENING, DECEMBER 29

8:00 P.M. *Symposium: The Scientist in American Society, Part II.* Arranged by a subcommittee of the AAAS Symposium Committee: Charles D. Coryell, Chairman, P.M. Morse, V.F. Weisskopf, Massachusetts Institute of Technology, and Bart J. Bok, Harvard University.

Edward U. Condon, Presiding

Papers 1. The Need for and the Production of Scientists
Harold C. Urey, University of Chicago.

2. Scientists and Other Citizens
Gerald Piel, The Scientific American.

3. The Legal Basis for Intellectual Freedom
Mark Dewolfe Howe, Harvard University

4. American Scientists and Public Affairs Since 1945
Edwin C. Kemble, Harvard University

Discussion: led by *Edward U. Condon*

WEDNESDAY MORNING, DECEMBER 30

9:30 A.M. *Symposium*: *The Metropolis*.
Joint session of AAAS Section E-Geology and Geography, the New England Division of the Association of American Geographers, and AAAS Section K. Arranged by Victor Roterus, Area Development Division, Department of Commerce.

Victor Roterus, Presiding

Papers: 1. Functions of the Metropolis
Lloyd Rodwin, Massachusetts Institute of Technology

2. Distribution of the Commercial Function in the Metropolis
William Applebaum and *Howard Green*, Stop and Shop, Inc. Boston

3. Residential Patterns of the Metropolis
Wayne F. Daughterty, Bureau of the Census

4. Spacing of New Growth
John T. Howard, Massachusetts Institute of Technology

Disscussants: *Prof. John Gaus*, Harvard and *Prof. Raymond Murphy*, Clark University.

WEDNESDAY AFTERNOON, DECEMBER 30

2:00 P.M. *Symposium: A Scientific Approach to the Problems of Delinquency*. Joint Session of the Society for the Advancement of Criminology and AAAS Section K—Social and Economic Sciences. Arranged by Donal E. J. MacNamara, New York University.

Donal E. J. MacNamara, Presiding

1. *James Brennan*, Director of Research, Juvenile Aid Bureau, New York City Police Department
2. *Melitta Schmiderberg*, Executive Chairman, Association for Psychiatric Treatment of Offenders
3. *Marcel Frym*, Director of Criminological Research, Hacher Psychiatric Foundation
4. *Colonel Mahmoud El Sebai*, Inspector of Police, Egypt
5. *Richard O. Arther*, John Reid Associates—The Use of the Lie Detector in Questioning Juveniles

Following the above joint conference came the newsletters of March 8 and August 3, 1954.

To Regional Researchers: March 8, 1954, Center for Urban and Regional Studies, Cambridge, MA

The American Association of Geographers will be holding its 50th anniversary meeting in Philadelphia, April 12–14, 1954 at the Hotel Penn Sherwood and Dietrich Hall, University of Pennsylvania. There are two sessions and several papers in other sessions which should be of interest to regional researchers.

A session on *Urban Geography* is scheduled for 9 A.M., April 14, Dietrich Hall, Robert E. Dickinson (Syracuse University) will be presiding. The papers and discussants are as follows:

Papers:

1. Recent Developments in the Economic Base Concept of Urban Growth and the Sector Theory of Neighborhood Growth
   Homer Hoyt, Homer Hoyt Associates

   Discussants: *John W. Alexander* (University of Wisconsin); *Edward L. Ullman* (University of Washington); W.*L.C. Wheaton* (University of Pennsylvania)

2. Urban Population Trends and Our Future Metropolitan Cities
   Robert C. Klove, Bureau of the Census

   Discussants: *Herlin G. Loomer* (Philadelphia City Planning Commission); *Raymond E. Murphy* (Clark University)

3. Industrial Location and Community Development in Metropolitan Areas
   Victor Roterus, Department of Commerce

Discussants: *Chauncy D. Harris* (University of Chicago); *Rita D. Kaunitz* (United Nations Secretariat); *Robert B. Mitchell* (University of Pennsylvania)

4. Accessibility as a Measure of the Extent of Urban Nodality in the Chicago Region
Harold M. Mayer, University of Chicago

Discussants: *John E. Bush* (Rutgers University); *Allen K. Philbrick* (University of Chicago)

A session on *Theory in Economic Geography* is scheduled for 1:30 p.m., April 14, Dietrich Hall. This session is of an interdisciplinary character. It has grown out of our previous informal sessions and has been arranged particularly to encourage the interaction of economists and geographers. Robert S. Platt (geographer, University of Chicago) will preside. The papers and discussants will be as follows:

Papers:

1. Theoretical Approaches to Spatial Analysis
W. Isard, economist
2. Substitution Analysis and the Location of the Petrochemical Industry
Eugene W. Schooler, economist, M.I.T.
3. Further Reflections on the Use of Input-Output Analysis in the Study of the Impact of Steel on the Delaware Valley
Robert E. Kuenne, economist, Harvard
4. Gravity Models and Supermarket Trading Areas
Howard L. Green, geographer, Stop and Shop, Inc.

Discussants: *Thomas R. Smith* (geographer, University of Kansas); *Edward L. Ullman* (geographer, University of Washington); *Harold H. McCarty* (geographer, University of Iowa)

Among other papers which should be of interest are: (1) "Geography as Applied to Railroad Industrial Development" by Wilbur R. Lamb (Chesapeake and Ohio Railway Company), April 12, Dietrich Hall; (2) "Applied Geography in the Industrial Development Program of the New York State Department of Commerce", by Donald M. Roznowski (Department of Commerce, State of New York), the same place and evening; (3) "Applications of Geography in Marketing Research" by Alan S. Gardner (Kaiser Corporation), the same place and evening; (4) "Geography as Spatial Interaction: Some American Examples" by Edward L. Ullman (University of Washington), April 13, Hotel Penn-Sherwood; (5) "Integration of the Use of Irrigated and Non-Irrigated Land in the Intermont West" by Wesley C. Calef (University of Chicago), April 14, Dietrich Hall; and (6) "Water Problems of the Trans-Rocky West" by Martin R. Huberty (U.C.L.A.).

A number of regional researchers have requested a reprint of the study, "The Impact of Steel upon the Greater New York–Philadelphia Industrial Region: A Study in Agglomeration Projection." They, and a few others whom I think will be interested, will find one enclosed along with a reprint of "The Economic Base and Structure of the Urban Metropolitan Region." This latter reprint in an over-simplified, non-technical fashion develops the approach used in the former. If any one who has requested these reprints does not receive them, please let me know.

Walter Isard

3 The Evolution of the Designations: Regional Science, Regional Science Association and the Field of Regional Science

As already noted, the group of regional researchers who came together in 1950 and earlier years were primarily regional economists. The signal December 1950 meeting was in effect an appendage to the annual convention of the American Economics Association. Only a few geographers, engineers and city planners were present. Then the major concerns of the group were: (1) increasing the stock of available relevant data; (2) improved data processing; and (3) identifying types of meaningful regions for which to attack diverse regional problems. The first major group effort (unsuccessful) was concentrated on having the Social Science Research Council support a Committee on Regional Economic Studies. All the members of the Working Committee set up for this purpose were economists.

A second meeting of regional researchers was again an appendage to a convention of an economics association, the Midwest. It was largely of regional economists, and data problems and the definition and identification of regions were once more the dominant concerns. However, soon thereafter we realized that more than economic factors were involved in attacking regional problems and development. There then followed meetings with the American Sociological Society, American Institute of Planners, the Association of American Geographers and the American Association for the Advancement of Science in order to develop multidisciplinary approaches in regional research. As a consequence, the notion of setting up a Regional Economics Association, which was informally tossed out for discussion at the December 1950 meetings and quickly judged to be premature, was discarded. While these multidisciplinary meetings were going on, which were unique in that they were restricted to regions and their problems, another basic thrust on research development emerged. This pertained to the need to evolve conceptual frameworks, models and theories, not only for attacking regional problems, but also to relate to pure spatial processes and phenomena. Thus after the public recognition of the work of regional researchers at the joint sessions on New England's problems and its confidence building effect and, as noted below, after more formal organization became necessary in order to join with the Allied Social Science Association (ASSA) in effective programming and hotel arrangements at conventions, we faced the problem of identification as an Association or Society.

Meanwhile, after the December 1950 meetings much informal and casual discussion took place in the corridors of MIT and Harvard University and at social gatherings of regional researchers on what we should call ourselves. Some, like Vining, Moses, and at times, Stevens and Isard, were interested in designating ourselves more formally as Spatial Researchers, Spatial Analysts or Spatial Scientists. But we recognized that such a designation would be exceedingly confusing for the mid-1950s— years of rocket development and extraterrestrial exploration widely reported upon by the media and of exceedingly great interest to the public. Another designation extensively discussed was: Association of Regional Researchers. But after the April 1954 convention of the American Association of Geographers with whom we scheduled two major sessions on regional research, it became clear that many, if not most, of U.S.A. geographers claimed to be regional researchers, and rightly so. However, most of the work of the U.S. geographers was concerned with essential data collection, processing, classification, cartography and other basic descriptive studies. Such involved little of the formal analysis, models, hypothesis testing and theoretical structures which were becoming of increasing interest to our group of regional researchers and which were aimed to be applicable for regional policy and development. But since U.S.A. geographers were regional researchers too, the designation "Association of Regional Researchers" was judged to be inappropriate and misleading and would raise problems.

Another name tossed around was "Association for Regional Studies." However, in the early 1950s there existed many regional study groups such as Harvard's Russian Research Center, and Centers and Institutes on Latin America, South Asia, China, etc. At these centers and institutes historical, political and broad economic studies were conducted on large political blocs and nations which had little to do with the research questions of our group on urban areas and subnational regions.

Also, we could have called ourselves "Association of Regional Analysts," but we claimed to be more than analysts. While we were data collectors and processors, we were also explorers of new conceptual frameworks, models and were becoming involved in highly advanced and deep theoretical thinking. We felt we were scientists. Finally, some of us toyed with the notion of Association for Regional and Interregional Science or Analysis, but adding on "and Interregional" would make the title of our association too lengthy and clumsy. We chose the crisper title "Regional Science Association" and indeed at that time most of our research dealt with the analysis of single regions and their problems, and little with interregional analysis. Still other designations that were suggested were "Association of Areal Scientists" and "Association of Unified Social Scientists." They found little support. So we converged on the designations: Regional Science, Regional Science Association and Field of Regional Science.

4 The Formation of the Regional Science Association

While there had been both much casual and some serious talk about a possible association of regionally oriented scholars, the possibility of forming an association was first explicitly raised in the following newsletter.

Dear Regional Scientists: August 3, 1954, M.I.T., Cambridge, Mass.

First, let me mention a meeting which may be of interest to you. On September 8-10, the American Sociological Society convenes in Urbana, Illinois (University of Illinois). There is scheduled for the morning of Friday, September 10, the following session:

Regionalism

Chairman: *Howard W. Odum*, University of North Carolina

Papers:
1. Cultural Regionalism and Area Studies
   J. H. Steward, University of Illinois
2. Regionalism and Metropolitan Areas
   E. A. Engelbert, University of California
3. Regional and Folk Sociology in Japan
   D. E. Lindstrom, University of Illinois
4. Geographic Structure of Regionalism
   E. L. Ullman, University of Washington
5. Economic Structure of Regionalism
   Walter Isard, Massachusetts Institute of Technology
6. TVA's Contribution to Regional Balance
   William E. Cole, University of Tennessee

Now for the more important matter. On a number of occasions we have discussed in one way or another the feasibility of forming an association of individuals concerned with regional problems and methods of regional analysis. Recently, interest in the establishment of an association has been mounting. Consequently, on May 19, 1954, I wrote the following to members of our two working committees:

"Since you are a member of one of the informal working committees on regional research which were set up some years ago to furnish advice when called upon, I am now writing you.

It seems that a number of regional researchers have the feeling that some steps ought to be taken toward a more formal organization in order to maintain the vigor and productivity of our group. We have been getting together on numerous occasions in connection with the meetings of different scientific associations. Looking back over the years I think we can truly say that we have had fruitful interaction and exchange of ideas. But it now appears to some of us that we are ready for a deeper type of interaction and synthesis of ideas than can be afforded by the more or less casual and informal meetings that we have had. In short it seems that it might be desirable for us to arrange annual meetings devoted exclusively to regional analysis and problems which might continue for two or three days. Such meetings would furnish both a greater opportunity and incentive for members of the several disciplines to convene regularly and explore ideas and techniques on a more penetrating level than we have hitherto been able to do.

To arrange such meetings on a regular annual basis would probably require the formation of some kind of an association, perhaps an Association of Regional Scientists. This need not entail any marked increase in organizational work or in financial obligations, at least over the course of the next few years.

I am therefore writing you to obtain your reaction and advice on this matter. Do you think that regular annual meetings would be desirable? If so, do you think that an association might profitably be formed? If your answers are in the affirmative on both these questions what would you suggest as appropriate objectives and activities of an association. I might remind you at this point that our regional research group includes individuals who are interested in one or more of the five following areas of research:

1. Pure spatial theory; general location theory; location analysis for individual industries; etc.
2. Structure and function of regions within the U.S.; regional projection techniques; regional income analysis; interregional analysis and trade; resource development; etc.
3. Urban metropolitan regions, their changing spatial structure and function; transportation system analysis; population and land use projections; labor market analysis; etc.
4. Large world regions; the process of economic development and industrialization; the interaction of cultural, social, political, economic and geographic factors; the impact of technology upon resource use; the urbanization process; etc.
5. Interdisciplinary analysis, the region being viewed as an ideal focal point because its problems require contributions from the several

social sciences and forces at least some integration of analytical techniques.

(The above is far from a complete statement of our areas of interest; and, as you well know, all five of these areas overlap in a major way.)"

In addition to writing to members of our working committees I have spoken about this matter with a number of other individuals. Thus far, except for two individuals every one has reacted positively to the holding of regular annual meetings, and to the formation of an association.

Accordingly I am proceeding to arrange a set of high-level sessions devoted to regional analysis to be held December 27-30, 1954, in Detroit. As I have already written to several individuals, it seems to me that it is best to begin by holding meetings along with an annual convention of a well-established scientific association at some central point in the United States. The A.A.G. (geographers) meets next year at Memphis; this is too far off the beaten track. The A.I.P. (planners) is also holding meetings next spring at an inconvenient place. The A.S.S. (sociologists) are holding their meeting in Urbana this September, but this is too soon for me to organize effectively a high quality program. This leaves the A.E.A. (economists) which is convening between Christmas and New Year in Detroit. It is the opinion of several with whom I have spoken that, in view of the relatively heavy concentration of regional scientists in the Detroit/Chicago urban areas, this is probably the best occasion at which to hold the first set of meetings.

In several weeks I hope to be able to write you again to give you more details about our plans for the meetings, topics to be discussed and participants. I shall of course welcome any suggestions and reactions you may have. With best regards,

Sincerely, Walter Isard

Given the decision to hold a set of high-level sessions at the time, December 27–30, 1954, of the American Economics Association convention and close by its meeting place we selected Hotel Teller as our venue. The next newsletter (September 30, 1954) presents some preliminary details about Hotel Tuller and sessions being scheduled. Also, it includes a call for papers and makes clear that the set of meetings being arranged is only "a purely exploratory step in considering the feasibility of an association of regional scientists." It concludes: "If we do not achieve a fruitful and profitable interchange of ideas, and if participants do not express a strong desire to continue annual meetings, we should, of course, drop the matter of an association."

Dear Regional Scientists: September 30, 1954, MIT, Cambridge, Mass.

I now wish to bring you up to date on the details of the regional sessions for December 27-30, Detroit.

Arrangements have been made to use the Hotel Tuller, Detroit, as our headquarters. All our sessions will be held in the Sky Room of the Hotel Tuller, except possibly for two sessions, one of which is sponsored by the American Economic Association, and the other by the Econometric Society. These may, or may not be held in the Sky Room of the Hotel Tuller. In any case, we will either formally or informally be joint sponsors of these two sessions.

The Hotel Tuller has been chosen for our headquarters primarily for two reasons. It is directly across the street from the Hotel Statler at which the American Economic Association will be holding its meetings. Its charge for rooms is much more reasonable than that of the Hotel Statler, although, of course, it does not provide the luxuries of the Hotel Statler. (Several persons have judged the facilities of the Hotel Tuller to be adequate for our purposes.) Enclosed is a reservation card. If you plan to stay at the Hotel Tuller, please make your reservation early.

The program, as currently planned, will consist of five main sessions and several other sessions built around shorter and contributed papers. Two sessions are in the final stages of organization. These are the ones which will be sponsored by the American Economic Association and the Econometric Society.

Many times individuals have expressed the view that researchers, particularly younger persons, who do have significant findings or methods upon which to report, often do not have an opportunity to do so, particularly when a set of meetings is completely organized by a relatively few persons. Accordingly, our schedule of sessions is so arranged that at least three to four sessions can be devoted to shorter papers, some of which may be invited and some of which can be contributed. Any person desiring to contribute a paper should write me directly, and, if possible, send along an abstract or a copy of his proposed paper. I have asked Leon Moses of Harvard University (economics) and Howard Greene of Stop and Shop (geography) to assist in judging whether or not proposed contributed papers are worthy of inclusion in our program.

There is one final point. It was not clearly indicated in the previous newsletter that, upon the advice of our working committees, I have arranged this set of meetings as a purely exploratory step in considering the feasibility of an association of regional scientists. Many persons have felt that the formation of either an informal or formal association would be desirable. The real test, however, will come in terms of demonstrated interest and participation in meetings that might be arranged. It seems to me that only after we have met and had an opportunity to obtain the deeper interaction and synthesis of ideas which we anticipate should we make any firm decision on the formation of an association. And we must make this decision bearing in mind the number and character of existing organizations with which each of us has affiliations. Therefore, I have not set up any formal organizational structure for these meetings. However, in order to arrange a headquarters hotel, an adequate meeting room for our sessions, and an appropriate listing of our sessions in the general program of the Allied Social Science Associations, I have found it necessary to speak of ourselves as an Association of Regional Scientists. This is only a tentative step. If we do not achieve a fruitful and profitable interchange of ideas, and if participants do not express a strong desire to continue annual meetings, we should, of course, drop the matter of an association.

With best wishes, Sincerely, Walter Isard

The November 10 newsletter presented a preliminary program; the final program appears in the following newsletter.

Dear Regional Scientists: December 9, 1954, MIT, Cambridge, Mass.

Enclosed with this newsletter is a copy of a paper by John Krutilla on "Criteria for Evaluating Regional Development Programs" which will be presented at one of our regional sessions. I am circulating this paper in advance of the meeting in order to foster fruitful discussion. The paper is preliminary. Next year, I hope to be able to circulate some of the better papers delivered at the meetings.

The program as it now stands is listed below. There will undoubtedly be some minor changes. You will note that all the meetings are being held in the Sky Room, Hotel Tuller except for the session scheduled jointly with the American Economic Association at 2:30 p.m., Tuesday, December 28. This session will be held in the Hotel Statler.

Program of the Association of Regional Scientists
Hotel Tuller

MONDAY, DECEMBER 27, 1954

2:30 P.M. Subject: *Regional Economic Analysis*
(joint with Econometric Society)

Chairman: *Edgar M. Hoover*, Washington D.C.

Papers:
1. Industrial Complex Analysis and Regional Development with particular reference to Puerto Rico
Walter Isard and *Thomas Vietorisz*, Massachusetts Institute of Technology
2. Estimation of the Balance of Payments of the Southeastern United States in 1950
Gloria Hile, Standard Oil of New Jersey

Discussants: *Julius Margolis*, Stanford University; *Nathaniel Wollman*, University of New Mexico; *Britton Harris*, University of Pennsylvania

---

8:00 P.M.

Chairman: *Richard Alt*, Arthur D. Little, Inc.

Papers:
1. The Role of Diversification in Local Economic Development
Allan Rodgers, The Pennsylvania State University
2. Remarks on Natural Resources Research and Regional Analysis
E. S. Wengert, University of Oregon
3. The Impact of Urban-Industrial Development on Agriculture in the Tennessee Valley and the Southeast
Vernon W. Ruttan, T.V.A.
4. Some Applications of Thünen's Model in Regional Analysis
Alexander Melamid, New School of Social Research
5. Some Reflections on Lösch's Theory of Location
Martin Beckman, Cowles Commission for Research in Economics

TUESDAY, DECEMBER 28, 1954

9:30 A.M. Subject: *Regional Analysis—I*

Chairman: *Amos H. Hawley*, University of Michigan

Papers:
1. Measured Urban Spatial Interaction as a Tool for Describing Metropolitan Regions
   J. Douglas Carroll, Detroit Metropolitan Area Traffic Study
2. Basic Elements of a Regional Resource Development Program
   Joseph L. Fisher, Resources for the Future, Inc.

Discussants: *Alfred C. Neal*, Federal Reserve Bank of Boston; *John W. Hyde*, University of Michigan; *Edgar S. Dunn*, University of Florida

---

2:30 P.M. Subject: *Regional Economics* (joint with the American Economic Association)

Chairman: *Walter Isard*, Massachusetts Institute of Technology

Papers:
1. Criteria for Evaluating Regional Development Programs
   John V. Krutilla, T.V.A.
2. Regional Reaction Paths to Changes in Economic Activity
   Frederick T. Moore, RAND Corporation

Discussants: *Philip Neff*, University of California at Los Angeles; *Leon Moses*, Harvard Economic Research Project; *Frank A. Hanna*, Duke University

---

8:00 P.M. Subject: *Regional Research: Selected Topics*

Chairman: *George H. Ellis*, Federal Reserve Bank of Boston

Papers:
1. A Method of Determining Incomes and their Variations in Small Regions
   Charles M. Tiebout, Northwestern University
2. The Role of Labor Mobility in Regional Growth
   William H. Miernyk, Northwestern University
3. Tax Incidence, Interdependence, and Regional Development
   James Jones and *Robert Kavesh*, Dartmouth College
4. Urban and Inter-Urban Economic Equilibrium
   Robert L. Steiner, Cincinnati, Ohio

WEDNESDAY, DECEMBER 29, 1954

9:30 A.M. Subject: *Regional Analysis—II*

Chairman: *Robert D. Calkins*, The Brookings Institute

Papers:

1. The Spatial Structure of an Economic System
Rutledge Vining, University of Virginia

2. Regional Analysis and Public Policy
Albert Lepawsky, University of California

3. Howard W. Odum, the "New Southern Regions", and Regionalism
Rupert B. Vance and *George Simpson*, University of North Carolina.

Discussants: *Harvey Perloff*, University of Chicago; *Victor Roterus*, U.S. Department of Commerce; *Frank Boddy*, University of Minnesota

---

2:00 P.M. Subject: *Regional Analysis—III*

Chairman: *Robert S. Platt*, University of Chicago

Papers:

1. The Need for an International System of Regions and Subregions
Donald J. Bogue, Scripps Foundation for Research in Population Problems

2. Spatial Interaction Analysis
Edward L. Ullman, University of Washington

3. Planned Decentralization and Regional Development with Special Reference to the British New Towns
Lloyd Rodwin, Massachusetts Institute of Technology

Discussants: *Seymour E. Harris*, Harvard University; *Edward A. Ackerman*, T.V.A.

5:00 P.M. *Business Meeting*, Association of Regional Scientists

---

8:00 P.M. Subject: *Regional Research; Selected Topics*

Chairman: *Louis B. Wetmore*, Massachusetts Institute of Technology

Papers:

1. The Spatial Impact of Transport Media
William L. Garrison, The University of Washington

2. Some Aspects of Regional Multiplier Analysis
   Lawrence E. Fouraker, The Pennsylvania State University
3. Economies of Scale and Regional Development
   J. A. Guthrie, State College of Washington
4. Fluctuations in Urban Growth: Some Comparative Data and Analysis
   Robert Williams, University of California at Los Angeles

In the days before the business meeting at Hotel Tuller, there was received many responses to the idea of organization. Some strongly opposed the formation of a new organization. The best stated and well thought-out one was that of Professor Derwent Whittlesey of Harvard University. In his letter of August 24, 1954, he wrote:

Dear Isard:

Thank you for your mimeographed information about the plans for the study of regions. It is one more evidence that a good many people have been interested in regions for a long time, and now others are discovering the importance of dividing the earth regionally.

I am enthusiastic about meetings such as that at the University of Illinois on September 10th. It is particularly good to note that the conference, although staged by the American Sociological Society, will listen to contributions by others than sociologists. No doubt a number of disciplines will in future plan special programs in which the regional aspects of their subjects are emphasized.

I cannot be at all enthusiastic about your suggestion that a new association be formed for the express purpose of studying regions and regionalism. I very much doubt if you will find wide enough interest to warrant forming such a group. In my opinion there are too many organizations already. By supporting many we divide and reduce our strength. As knowledge proliferates, it is inevitable that people who work in border fields will wish to exchange ideas from time to time. In some instances it may be appropriate to set up a border field as a discipline on its own platform. I do not think regional study is a border field in the sense of biochemistry or political geography. (Even if it is, I think a new organization is a mistake.) To be a viable discipline, I believe a subject must view a segment of the circle of knowledge in a way not already being handled by existing disciplines.

Now, regional study is the core of geography, and always has been. Every session of geographers is primarily concerned with some sort of region, and sessions on the theory of regional study would welcome the approaches of your economists and students of other social and natural sciences. Put in practical terms, there already exists a forum for discussing regions in the broadest possible manner, orally and in print. Any serious student can readily become an active member of this existing "Association of American Regionalists (Geographers)," and thus combine the strength of all persons interested.

If a new organization were formed, by far the largest group interested would be geographers. However, it is my guess that the number of members would be few. Thus the new organization would lack the support of the largest group by far that give regional interests high priority. You must know how precarious is the infancy of any scholarly organization. The scraps of soon-dead "organs" that lie unremembered on library shelves and elude the research of even painstaking scholars are witnesses of the difficulty in bringing to maturity any such association.

I should be eager to work for full recognition by the A.A.G. of the regional contributions from cognate disciplines. It occurs to me that *The Professional Geographer* could very properly publish advance notices of all sessions on regional study, thus performing at no extra cost one of the useful tasks that you have yourself undertaken. The *Annals* of A.A.G. would certainly be hospitable to well-studied manuscripts on regional study, whatever the approach and the character of the material.

The Response by Isard was:

Dear Whit:

Thanks ever so much for your letter of August 24. Your views are extremely valuable to me in weighing alternative courses of action.

Actually, my news letter of August 3 may be somewhat misleading. No final decision has yet been made to form an association of regional scientists. All I am doing is arranging a set of meetings in Detroit in order to begin exploring more fully this possibility. As I have indicated, many persons have thought the formation of an association desirable. But I firmly believe that unless there is deep interest and widespread participation it would be unwise to form any kind of an association. For the present, no formalities are envisaged.

The core of the problem is how can we obtain deeper interaction and synthesis of ideas than we have hitherto had among various social scientists concerned with regional problems and analysis. We certainly cannot achieve it by having meetings with established associations as we have had in the past. At the A.A.G. meetings in Philadelphia last year, and at Cleveland in the previous year, we have had sessions which involved regional analysis on a level which appeals to the various members of our regional researchers' group. Unfortunately, at both those sessions only very few non-geographers found it expedient to attend. There were not enough sessions on this level of analysis to make it worth while for non-geographers to journey a long distance (except for some non-geographers in the local area.) Perhaps it may be that the sessions at Detroit will not attract a sufficient number of persons. If they do not, then the answer is clear that the formation of an association should not be attempted. If they do, and if there is fruitful and profitable interchange of ideas, and if the participants express a strong desire to continue annual meetings, then I do not see why annual meetings should not be planned. Further, if the participants express a strong desire to band together into an organization, then I think that their wish should be given serious consideration, especially if among them there are those who are willing to bear the financial and administrative burden.

In exploring this matter there are several other points to keep in mind. One, the group of regional researchers now numbers about 200 persons, most of whom are quite active. Secondly, it is not easy to get publication in geographic journals of manuscripts which would interest at least one half the members of our regional research group, but which, because of their mathematical content, would interest only a handful of geographers. Further, about three-quarters of the members of the regional researchers' group are not geographers, and do not receive *The Professional Geographer;* hence *The Professional Geographer* can not effectively keep the various members of our group informed on various developments. Thirdly, it is not easy to arrange sessions at the annual meetings of established professional organizations. For example, the request that at the Memphis meetings of the A.A.G. a special session be set up on the more theoretical aspects of economic geography and of urban and regional analysis was more or less turned down. I think there was good reason for this; namely, that the appeal to geographers would be limited, and that attendance of non-geographers at this session, especially as it is at Memphis, would also be sparse.

Thus, you see that there are a number of problems confronting those of us who seek to get together for a deep, penetrating discussion of materials on a fairly abstract level. Remember, we are merely exploring the matter of an association. What happens will largely depend upon the desires, interest and participation of those engaged in regional research of all kinds.

In general, most geographers were not in favor of setting up an association of regional scientists. However some of the leading young ones, like Ullman and Ackerman, were supportive; for example, Ackerman wrote: "I am willing to go along with the association you suggest because it is so close to the heart of my interests. Those interests, I suppose, are touched on closely in the AAG, but the geographers still have been a little clannish in their approach to regions. Men from other fields, I know, would find no comfort as a satellite unit of AAG."

Those from other fields frequently stated that we already have too many organizations; we do not need one more. From sociology and demography, Bogue wrote in his July 23 letter:

"Although I have been very interested and educated by the series of sessions and releases which you have sponsored over the years, I think we should go very slowly in forming a separate professional society. From my point of view it would be much better to get established as a working branch of one of the professional societies or to continue as we have in the past. As it now stands there are far too many associations and conventions to attend, and I think little would be gained by setting up a formal association with respect to regionalism. The subjects which you outlined in your May 19 memorandum are not sufficiently unique to constitute a separate sphere of interest. Almost everyone interested in these subjects has another professional field which he uses as a point of reference. I strongly doubt whether you can make regionalism as such a point of reference and retain enough members to keep a professional society alive.

The American Geographical Society has a very broad and tolerant point of view and could very well provide the haven which you seek. The sociologists also would welcome you."

Odum, the distinguished leader of regionalism, wrote "As a rule I have been skeptical about forming new associations, but as subsequent events continue to emphasize the significance of such a program, I should be very happy indeed to co-operate in any way that I can" and very wisely stated: "I am glad to see that you are including world regions, and am inclined to think that we may want a scientific and theoretical item, perhaps Number three, featuring regionalism as cultural functioning as well as structure within an areal framework."

Other sociologists were positive. Kingsley Davis wrote: The formation of an association devoted to regional studies is an excellent idea. Vance, in support, stated "I do hope that you will find someone to give a good account of sociology and its techniques in this collaboration. This can be a most stimulating group." Blackwell felt that organization is a good idea to try. Whitney wrote in a similar vein. While he regarded the interdisciplinary approach as being dubious in many respects, he felt that annual meetings as real working conferences providing stimulation from direct contact of diverse scholars is fundamental and should be the basis for organization.

Among planners, Blucher was opposed to another organization, but Perloff was "very enthusiastic about the whole idea of an association." Mitchell, Adams, Rodwin and others, were in general quite positive. Among economists, largely regional ones whose reaction was solicited, there were a number of enthusiastic responses. These responses were a reflection of the need of regional economists to find opportunities for presenting their findings at meetings and to have professional recognition. This need (as well as the continuing spatial bias of economics) was pointed to in the response of Seymour Harris to organization: "This, at least, will assure you some time and space at the annual meeting of the American Economic Association."

(Parenthetically it should be noted that in the early '50s the persistence of spatial bias was also reflected in the constant rejection of proposed books (by Isard) on *Location and Space Economy* and *Methods of Regional Analysis*, these negative responses being by the Columbia University Press, McGraw Hill, Prentice-Hall, MacMillan, the University of Chicago Press, to name several. The referees engaged by these presses were of course balanced and well acquainted with standard economic doctrine. Samuelson, for one, regarded the Location and Space-Economy book as much too formal to realize sales of 1,000 copies considered necessary by a publishing house. However, it did turn out that at last both proposals found acceptance by the editor of the MIT Press who at the time of the above acceptance (June 1954) had been unsuccessful in obtaining any book proposal of quality. This was so since the MIT Press had been recently reorganized to be a book publisher, rather than being involved only in putting out mimeographed and other materials for classroom use. It turned out that each of the two books accepted had sales of 25,000 or more and were, in addition, translated in several languages.[11]

Returning to the meeting at the Hotel Tuller, it turned out to be extremely stimulating and intellectually provocative. As reported below in the minutes of the business meeting a motion to organize formally was unanimously approved by the

---

[11] Another instance of bias in economics was the need to avoid use of the term gravity model in papers submitted to journals for publication, the notion of such a model considered by some to be quackery. For example, for the acceptance of the two-part 1954 *Quarterly Journal of Economics* article by Isard and Peck, the word gravity had to be deleted and in its place, the much more acceptable term income potential had to be used. (Strictly speaking, a concept of mutual income potential should have been employed to parallel the mutual energy concept of physics, but such would have suggested overtones of the gravity model.) It should be noted that in the early 1960s Jan Tinbergen with some of his students asked to have lunch with Isard during one of the European Regional Science conferences. Subsequently, Tinbergen became the first leading economist to use a gravity model formulation in his thinking. Already by 1960 in the *Methods of Regional Analysis* book, the gravity model had become standard in regional science literature.

sixty or so scholars in attendance. Further, it was decided to convene again at the time (December 1955) and place (New York) where the American Economic Association was planning to meet and where also the A.A.A.S and the American Statistical Association were to convene. Isard was authorized to set up a committee to draft a constitution for the organization.

Minutes of Business Meeting—Association of Regional Scientists
Hotel Tuller, Detroit, December 29, 1954

Attendance: approximately 60

Isard:

Basic question: Should we organize to hold annual meetings on regional analysis?

Gibelman:

Move that we formally organize as the American Association of Regional Scientists.

Garrison:

Substitute motion; omit name. Let a committee decide on a name.

Miernyk:

Strong rationale for this group. Need interdisciplinary organization for cross-fertilization of ideas. Would like to see papers published as proceedings.

Reid:

Would like to see the group formalized, particularly since it offers a focus for both practitioners and academicians.

Isard:

In correspondence, several persons have voiced opposition to a proliferation of organizations. We should seriously discuss this matter.

Chapin:

From experience at University of North Carolina, many problems in achieving effective interdisciplinary approach. Tough sledding should be expected at start, certainly until the problem of semantics is conquered. This, however, should not be interpreted as indicating that we should not proceed with organization.

General discussion followed. The question was raised about whether or not we could continue as an informal group. It was reported that the problem of organizing a set of meetings at a given time and place, especially in conjunction with other social science organizations, was made much more difficult by not having organizational status. Isard indicated all we need is a skeleton type of organization.

E. Harris:

In the field of regional research there is no journal. Therefore need to have circulated various memoranda and papers on current and regional research.

Jaidiaw:

Can the papers presented be duplicated for circulation?

Isard:

If we are to circulate papers we should perhaps circulate only the better ones in order to minimize expense. In such a case we should have an editorial committee. However, this would increase the extent of our organization.

Kavesh:

The group should be formalized sufficiently so that we are recognized as an organization. This is an important practical matter for those who must justify their travel expense to universities or business firms.

Wolff:

Would seem reasonable for all to make some contribution to defray whatever expenses are incurred.

Isard:

Unless we move into publication of the papers, it does not seem necessary at this point to have a budget.

Rodwin:

Want to return to the question of organization. Let us first decide if we want to organize formally. If we do, then set up a committee to handle the questions of budget, circulation of paper, and details of organization.

* * *

General discussion followed. The substitute motion was seconded. It was approved unanimously.

There followed discussion on the size, geographic distribution and the various disciplines to be represented by members of the proposed committee. It was felt that, among others, the committee should have two responsibilities:

1. To draft objectives of organization.
2. To draft alternative structures of organization.

These drafts should be sent to persons present and others on the mailing list for full discussion at the next annual meeting. There was also strong feeling that, if possible, the committee should have a broad geographic representation. If travel expense precludes a full meeting of the members of the committee, it was suggested that those on the Eastern Seaboard meet as a subcommittee and, by correspondence, incorporate in its proposals suggestions of members in other regions. After many proposed motions, the final amended motion was that W. Isard be given full freedom in selecting committee members and in determining the size and geographic representation of the committee, keeping in mind, however, the various suggestions made. This was passed unanimously.

Discussion followed on the date and place of the next annual meeting. It was felt that we ought to meet in conjunction with at least one other social science organization. Since the next annual meetings of the Association of American Geographers and the American Institute of Planners are scheduled to take place in Memphis, Tennessee, and in Kansas City, meeting with either or these two was ruled out. A meeting with the American Sociological Society in Washington, D.C. in September was considered more desirable. However, the decision was made to convene in New York during the last week in December, together with the A.A.A.S., the American Statistical Association, and the American Economic Association. It was urged that joint meetings be arranged with these Associations, and in particular with the Geography section of the A.A.A.S. Meeting was adjourned. Robert A. Kavesh, Secretary.

The next newsletter of March 14,1955, with which were enclosed the minutes of 1954 business meeting, records organizational developments and announces new meetings in standard fashion. The newsletter of July 28 reports steps taken for the formation of an association and the newsletters of September 20 and December 12 are concerned with and announce the 1955 December program in which younger scholars are encouraged to submit papers for consideration for presentation.

Dear Regional Scientists: March 14, 1955, MIT, Cambridge, Mass.

Enclosed are the minutes of our business meeting at Detroit. We are proceeding with the arrangement of a set of meetings to be held December 28–30 in New York, and which I hope will be even better than those at Detroit. Several of our sessions will be jointly held with other associations meeting at the same time and place.

Also, I am planning to circulate a number of the Detroit papers, probably during the summer when secretarial assistance will be available. We have not yet decided on the exact form in which these papers will be circulated.

Already Bob Kavesh has drawn up a first rough draft of an organization for Regional Scientists for discussion by our organizing committee. You shall receive, well in advance of our December meetings, a copy of the committee's proposals on organization.

Of interest to some is the Regional Income Conference sponsored by the National Bureau of Economic Research, Duke University, June 17–18, 1955. The following is the *tentative agenda* (discussants of the papers are not listed):

Papers:

1. Value of the Regional Approach in Economic Analysis
   Walter Isard, M.I.T.
2. Problems of Assessing Regional Economic Progress
   Harvey Perloff, University of Chicago
3. Analysis of Inter-State Income Differences: Theory and Procedure
   Frank Hanna, Duke University
4. Inter-Regional Differentials in Real Income
   Dorothy Brady and *Abner Hurwitz*, B.L.S.
5. Measurement of Regional Income and Product
   Charles F. Schwartz, Office of Business Economics
6. The Geographic Area in the Study of Economic Problems
   Morris B. Ullman and *Robert Klove*, Bureau of the Census
7. City-Size and Income, 1949
   Edward Mansfield, Duke University

In addition to the foregoing "general" papers, a round table on county income problems has been arranged for the afternoon of June 18. It will center around four topics, as listed below.

Papers:

1. Conceptual Issues of Income Estimation for Local Areas
   Werner Hochwald, Washington University and St. Louis Federal Reserve Bank
2. Development of Inter-Censal Production Estimates for Local Areas
   Henry Shryock, Bureau of the Census
3. Measurement of Agricultural Net Income by Counties
   John Fulmer, Emory University
4. Uses of Short-Cut Methods in Estimating Local Area Incomes
   Lorin A. Thompson, University of Virginia

Sincerely, Walter Isard

Dear Regional Scientists: July 28, 1955, MIT, Cambridge, Mass.

As authorized during the December business meeting I have set up an organizing committee whose members represent several disciplines and the various regions of the United States. The members are: Joseph L. Fisher, Associate Director, Resources for the Future, Inc., (Economics); Chauncy D. Harris, Dean, Division of Social Sciences, University of Chicago (Geography); Albert Lepawsky, Professor of Political Science, University of California; Rupert B. Vance, Kenan Professor of Sociology, University of North Carolina; and Louis B. Wetmore, Chairman, Department of City Planning, University of Illinois. In addition, Robert A. Kavesh serves as secretary, and I as chairman.

Instead of drafting alternative structures of organization, we considered it more desirable to develop a single form of organization which incorporates in a consistent framework the desirable features of various existing organizations. Accordingly, Mr. Kavesh drew up a first draft of a constitution. This first draft was then considerably modified and extended by the several members of the committee. The resulting statement is enclosed with this newsletter for your reaction and comment. The objectives of the association are briefly set forth in Article II.

It is the feeling of several members of the committee that if there are no major objections to this draft, i.e., if this draft meets with the general approval of our membership, we should not defer to the next business meeting the discussion of the proposed constitution. Rather, we should modify the proposed constitution in accordance with further suggestions received by September 15th and submit it to the membership for ratification.

* * *

Mr. Carrothers is busy editing the first volume of the Papers and Proceedings. We expect to distribute this volume in late summer or early fall.

Sincerely, Walter Isard

Dear Regional Scientist: September 20, 1955, MIT, Cambridge, Mass.

The program for the meetings of the Regional Science Association in New York, December 27-30 is coming along well. All the papers for the main sessions have been arranged. However, the structure of the program has not yet fully jelled. There will be altogether 5 or 6 main sessions, with several papers on regional theory, problems of urban-metropolitan regions, and problems of resource-type regions. We plan to publish the papers by photo offset or similar process. I expect that the sessions will be even more stimulating than those at Detroit.

Since I have not had an opportunity to extend invitations for discussants and chairman of all main sessions, the list of participants is incomplete. So far, the following have accepted invitations to participate in the main sessions:

Robert Calkins	Louis Wetmore	Morris Garnsey
Alfred Neal	Walter Isard	John T. Howard
Guy Freutel	E.T. Grether	Daniel Price
Reginald Isascs	Donald Bogue	Robert Kavesh
Chester Rapkin	Theodore R. Anderson	Richard Ratcliff
Stephen Jones	Joseph Fisher	James Jones
Nathaniel Wollman	Seymour Harris	Harvey Perloff
Stefan Robock	Vincent Ostrom	Preston James

As we recognized in arranging last year's meetings, researchers, particularly younger persons, who do have significant findings or methods upon which to report, often do not have an opportunity to do so, particularly when a set of meetings is completely organized by a relatively few persons. Accordingly, our schedule of sessions is so arranged that at least three to four sessions can be devoted to shorter papers, some of which may be invited and some of which can be contributed. Any person desiring to contribute a paper should write me directly, and, if possible, send along an abstract or a copy of his proposed paper.

Sincerely, Walter Isard

Dear Regional Scientist: December 12, 1955, MIT, Cambridge, Mass.

Enclosed is a copy of the revised program of the Regional Science Association, December 27-30. The rooms for most of the sessions have been changed. Please throw away the old program in order to avoid confusion.

Christmas greetings. Sincerely, Walter Isard

Program: Regional Science Association
Biltmore Hotel, New York, December 27–30, 1955

TUESDAY, DECEMBER 27

2:30 P.M. Subject: *Regional Science: Nature and Scope*

Chairman: *Robert D. Calkins*, Brookings Institute

Papers: 1. Regional Science and Regional Structure
Walter Isard, MIT

2. The Dimensions of Regional Science
Morris E. Garnsey, University of Colorado

Discussion: *Arthur A. Maass*, Harvard University; *Guy Freutel*, Federal Reserve Bank of St. Louis; *Daniel O. Price*, University of North Carolina; *Robert S. Platt*, University of Chicago

---

8:00 P.M. Subject: *Contributed Papers*

Chairman: *Louis B. Wetmore*, University of Illinois

Papers: 1. The Spatial and Interregional Framework of the United States Economy: an Historical Perspective
Douglass C. North, University of Washington

2. Location Factors in Synthetic Fibre Production
Joseph Airov, Emory University

3. Measuring the Economic Base
Charles L. Leven, Federal Reserve Bank of Chicago

4. Some Needed Improvements in Regional Income Estimation Procedures
John H. Cumberland, University of Maryland.

WEDNESDAY, DECEMBER 29

9:30 A.M. Subject: *Basic Interrelations In Regional Development*

Chairman: *Stephen B. Jones*, Yale University

Papers: 1. Development Problems in Light of Brazilian Experience
Stefan H. Robock, Banco do Nordeste do Brazil

2. Interrelations of Income and Industrial Structure
Harvey S. Perloff, Resources for the Future, Inc.

3. Political Dimensions of Regional Analysis
Vincent Ostrom, University of Oregon

Discussion: *Preston E. James*, Syracuse University; *Joseph Spengler*, Duke University; *Maynard Hufschmidt*, Harvard University.

2:30 P.M. Subject: *Basic Research Problems of the Urban-Metropolitan Region*

Chairman: *Robert B. Mitchell*, University of Pennsylvania

Main Speaker: *John T.* Howard, M.I.T.

Discussion: *Ernest M. Fisher*, Columbia University; *Reginald R.. Isasacs*, Harvard University; *Gerald Breese*, Princeton University; *Lyle C. Fitch*, N.Y. Division of Administration.

5:00–6:00 P.M. *Business Meeting*

8:00 P.M. Subject: *Contributed Papers*

Chairman: *Joseph L. Fisher*, Resources for the Future, Inc.

Papers:
1. Regional Advantage in Oil Refining
   Robert Lindsay, Federal Reserve Bank of N.Y.
2. Projecting Industrial Growth of Metropolitan Regions
   Britton Harris, University of Pennsylvania
3. A Conceptual Framework for the Analysis of Urban Development Policy
   George Duggar, University of California
4. Some Attempts at Estimating Parameters on Spatial Interactance
   William L. Garrison, University of Washington.

THURSDAY, DECEMBER 29

9:30 AM Subject: *Urban Growth and Development* (Joint Session: R.S.A., American Economic Association)

Chairman: *Walter Isard*, Mass. Institute of Technology

Papers:
1. The Spread of Cities
   Donald J. Bogue, University of Chicago
2. Changes within Cities
   Chester Rapkin, Columbia University

Discussion: *Richard U. Ratcliff*, University of Wisconsin; *Martin Meyerson*, University of Pennsylvania; *Joseph L. Fisher*, Resources for the Future, Inc.

2:30 P.M. Subject: *Statistical Investigations of Regional Structure and Reactions* (Joint Session: R.S.A., American Statistical Association)

Chairman: *Ewald T. Grether*, University of California

Papers:
1. Regional Variations in Money, Credit, and Interest Rates
   N. Wollman, University of New Mexico
2. Differential Regional Impact of Federal Expenditure
   Robert A. Kavesh and *James B. Jones*, Dartmouth College

Discussion: *Alfred C. Neal*, Federal Reserve Bank of Boston; *Seymour E. Harris*, Harvard University; *Harry Schwartz*, Federal Reserve Bank, San Francisco; *Paul B. Simpson*, University of Oregon

---

FRIDAY, DECEMBER 30

9:30 A.M. Subject: *Gravity and Potential Models*

Chairman: *William Applebaum*, Harvard University

Papers:
1. Potential Models and the Spatial Distribution of Population
   Theodore R. Anderson, Yale University
2. Market Potential Concept and the Analysis of Location
   Edgar S. Dunn, Jr., University of Florida

Discussion: *A. J. Jaffe*, Columbia University; *Howard L. Green*, Stop & Shop, Inc., *Gerald A. P. Carrothers*, M.I.T.

2:30 P.M. Subject: *Contributed Papers*

Chairman: *Leon N. Moses*, Harvard Economic Research Project

Papers:
1. Metropolitan Suburbs: A Descriptive Study
   Lee F. Schnore, Brown University
2. Factors in the Location of Administrative Offices, with Particular Reference to the San Francisco Bay Area
   Donald L. Foley, University of California
3. Hawaii, A Case Study in Regional Analysis and Area Development
   James H. Shoemaker, Bank of Hawaii
4. A Note on the Potential of Income and Population
   Martin J. Beckmann, Center for Advanced Study in the Behavioral Sciences

At the December 1955 business meeting of the Association (the minutes of which follow) the specifics of the first draft of the constitution were then discussed. It is again to be noted that in drafting the Constitution a first decision made by the working committee was to propose at the start a full constitution to be distributed to members for their reactions by communications and discussion at forthcoming business meetings. It was judged that it would be very difficult and time consuming to construct a constitution section by section since people would want to debate and vote on each section. Further, while the constitution would be a formal statement, the decision was taken to make the statement simple. Also, in order to insure the full independence of scholarly work, it was decided not to allow institutions to obtain membership, thereby to minimize the possibility of bias in the direction of research which often creeps in with financial support from institutions.

Also, it was noted that 200 copies of Volume I (1955) covering the December 1954 Papers and Proceedings of the Regional Science Association were prepared by having each paper-giver send 200 copies of his paper for editing in a neat and orderly fashion by Gerald A. P. Carrothers.

Minutes of Business Meeting—Regional Science Association, Hotel Biltmore, New York, December 28, 1955

Attendance: approximately 50

Walter Isard presided.

The minutes of last year's business meeting held at the Hotel Tuller in Detroit were read and approved. It was pointed out that the organization of the Association had been delegated to a committee to be appointed by Walter Isard—the reports of the committee to be referred to the entire membership for discussion and ratification. Members of the organization committee are: Rupert Vance, Chauncy Harris, Joseph Fisher, Albert Lepawsky, Louis Wetmore, Walter Isard (chairman), Robert Kavesh (secretary).

The chairman attributed the success of the 1955 *Papers and Proceedings* (Vol. I) to the cooperation of participants in the Detroit meetings and to the efforts of Gerald Carrothers. Unanimously Mr. Carrothers was reappointed editor for the 1956 *Papers and Proceedings.* In addition, since a large part of the financial transactions of the association arise in connection with the *Papers and Proceedings*, it was felt that the editor of the *Papers and Proceedings* should also serve as Treasurer *pro tem.* The association voted unanimously that Gerald Carrothers be appointed Treasurer and Messrs. Carrothers and Isard be empowered to sign checks for the account of the Regional Science Association, deposits of which are held at the Harvard Trust Company.

Re: next year's financial operations the chairman indicated that the organization might face a deficit because of the increased size of the mailing list and additional costs in printing the *Papers and Proceedings*. He suggested that all new members be asked to contribute one dollar. A substitute motion by Preston James, in which *all* members would be asked to contribute one dollar, was passed unanimously.

The Secretary analyzed the responses to the draft of the Constitution which had been circulated to the membership. The following suggestions were made: (1) a clause should be inserted in the Constitution stating that the RSA is a non-profit organization, and (2) a clause should be inserted to provide financial protection for officers.

The next matter discussed concerned the provisions of Article III, Section I, dealing with membership. Several persons urged that the Regional Science Association have a simple eligibility clause as do other organizations. Others pointed out that a young organization must be composed of interested persons and that others could qualify as "Associate Members" under Article III, Section II. After considerable discussion it was voted unanimously to abandon the two classes of membership and to liberalize the qualifications for membership.

Further discussion of the Constitution centered on the method of electing officers as stated in Article IV, Section II. No motions dealing with changes were passed.

The question was raised as to whether or not the adoption of the Constitution was in order at this time. Louis Wetmore suggested that ratification should await suitable modifications and that the purpose of the present discussion was to determine the general suitability of the document.

It was decided unanimously that the next steps should be:

1. Ratification of the Constitution (with revisions)
2. Appointment of a nominating committee, by the organization committee
3. Election of officers.

The next issue concerned a meeting place for 1956. It was decided to meet in Cleveland, December 27–30, 1956 in conjunction with the American Economic Association and other social science organizations.

The following resolution drafted by John Cumberland at the suggestion of Harvey Perloff was adopted unanimously:

> "Resolved that the Regional Science Association will support the Treasury Internal Revenue Service in its program to publish Statistics of Income for Metropolitan Areas. Further, that the Treasury be urged to explore the possibilities of extending this program to counties using refined statistical sampling techniques in conjunction with electronic tabulation to reduce current publication lags."
>
> The meeting adjourned at 6:15 P.M. with a rising vote of thanks to Local Arrangements chairman, Robert Lindsay and Robert A. Kavesh, Secretary.

Shortly after the 1955 business meeting a letter from Roy E. Bessey dated December 28, 1955 was received which raised some serious and meaningful questions about the objectives and structure of the new Association. They were as follows:

> As I have thought—or wondered—about the Association from time to time, a number of questions have come to mind. And the current program sort of points them up. Do you mind if I pass them along for what they may be worth in your own thinking?
>
> Is the membership and program too highly academic, theoretical, pure-science, cloistered? Do they reflect the other—the practical, the applied-science, the humanistic—sides well enough?
>
> Without any prejudice to the pure-science and its development, can the membership and program be more catholic? Bring in more different kinds of people and activity—especially those relating to the general practice of regional science and art?
>
> Should there be some more stress on the synthesis of the regional sciences—on regional science as an integrated whole—in relation to the coverage of the more specialized intra-disciplinary aspects? Morris' comprehensive and excellent piece helps, I think.
>
> Where do and where will the regional scientists ply their trade? (From the program it would appear that now it is mostly in the colleges, with some scattered exceptions in banks, and not in government at any level.)
>
> Does the program tend to be incestuous in the intellectual, academic family?
>
> Should aims and activities be oriented more toward the public, which is so deeply concerned with results, and toward the government which must implement most programs? Similarly, there should perhaps be more considera-

tion of regional science in transportation and distribution? Should the program be brought closer to the people who make the regional decisions—now generally without benefit of regional science?

In an unfolding program, how will the regional scientists look at the training of people to work with the planning, decision and action agencies? How will it look to the training of the applied-science people, the full-time, part-time and occasional practitioners in the field? How will the program inform the people who make plans and decisions of the very real need of regional and scientific approach and method? How will it get across—to the people and their representatives in government—some idea of the need?

Will the Association wait for a renaissance in applied regionalism—or should it help to bring it about? Can it help to bring about a recognition of the basic needs of regions as well as of the great need for regional science—theoretical and applied—in an increasingly complex, crowded, and rapidly changing world? Should the Association try to play a part in an awakening of government to the needs and values—and the essentiality—of a regional, scientific approach?

Is the Association interested in exporting some regional science? (It will find a more ready and willing market abroad than at home!)

Should regional science actively seek places to go, and how? Is its attitude in this regard passive or active? What is the Association's logical place in the development and use of regional science, as well as in the development of its theory and methodology?

The questions are somewhat overlapping. They could be boiled down to one question: of synthesis of disciplines and of theory and practice. On the other hand, they could well be made more detailed and specific.

Of course, little if any of the foregoing line of questioning is original. It's an old line, directed against a natural tendency to "in-grow" in intellectual, professional and similar activities—to talk and work and trade with those of like mind and interest. Whereas, the situation calls also for a good deal wider range of contact and effort. I noted—as we all did—some feeling on the part of others who reviewed preliminary outlines and material of the "Guide," that something was needed to make and present regional science as something more real and pragmatic and less mystical. Perhaps the U.N. people put this most strongly—they are looking for something both scientific and practical to put to use widely and rather quickly. Then, too, I got a rather similar feeling from reading Dwight Macdonald's articles in *The New Yorker* on the Ford Foundation. I don't know enough about the Foundation to know how much of the direct or implied criticism is justified or fully jus-

> tified, but out of some knowledge of the tendencies of people and organizations one can feel the sharpness of the points. One also gets a feeling of wide applicability and transferability about criticisms of this kind. Such organizations—and groups like the regional scientists—must be ever self-critical; they must keep their eyes constantly on the end purposes and uses and results; they must rise above institutional and professional limits in their thinking; they should speak a more universal or widely understandable language; they should foster individual work and thought for its higher potential peaks as well as group work for its wider spread and correlation; they shouldn't be excessively scientific in attitude; they should maintain and express imagination, enthusiasm, venturesomeness.
>
> I trust you'll forgive the "sermon"—I know it's unneeded, but I talk on these subjects on the slightest provocation. My excuse is that RSA is new and that critical thought on the part of all of us may be especially worth while at this time.

These questions, as raised by Roy E. Bessey, had of course been raised in one form or another by other scholars as well in the extensive back and forth discussion that had taken place ever since the idea of an association was first seriously considered in 1953. The reply by Isard to Bessey's letter was

> First let me tell you how much I appreciated having your very thoughtful comments. These are indeed extremely valuable for plotting a direction of any new organization. Your first point has already come up for considerable discussion. We are in the process of liberalizing the requirements for membership, so that the Association will not be too academic and theoretical and public administrators are tending to come in and nonprofessionals and businessmen are also about to join or joining. In short, the membership is coming to be quite diverse and yet I hope not so diverse as to lose a central focus. In this process the idea of synthesizing various types of materials that bear upon regions is high in the minds of all of us. I would judge that regional scientists will be employed in universities and banks, but also in government and business. The meetings of the Association emphasize current research, new ideas and new synthesis. It turns out as one would expect, that most of the research and synthetic work is going on at universities, and hence one finds that a greater part of the participants are from universities.
>
> As already indicated, the needs of governmental agencies are becoming increasingly appreciated in the organization. Considerable emphasis in research is on transportation and distribution, especially in the research on

gravity models. We have among us a large number of planners, but I do feel that it is not the role of the regional scientist to educate people and to put across ideas to people. This is more the function of the educator and we certainly don't want to spread regional science out so thin that it loses its driving force. We are starting to export regional science. We are organizing as an international association and already have some foreign people on our mailing list. I think these comments provide some sharp answers to your various questions. One must be careful not to be too ingrown, but at the same time not cover the water front.

The above letter and reply, as well as the discussion of the business meetings of 1954 and 1955, well define the general directions in which the Regional Science Association was to take in the years that followed.

The February 1956 newsletter announces informal and joint sessions at the annual conventions of the Association of American Geographers and the American Institute of Planners. The subsequent April 13 newsletter announces the establishment of the Ph.D. program in regional science which will be discussed at length below. This newsletter also reflects the concern of the Association and its members with nodal regions (urban and metropolitan areas) and their development as is recorded in the enclosed summary of discussion which took place. The May newsletter announces a joint meeting with the American Political Science Association and that the annual RSA convention will be held in Cleveland, Dec. 27-30. With the July 7 newsletter there is distributed a revised constitution of the Association, again subject to suggestions for change by members.

Nodal Regions: Summary of Joint Session of the Association of American Geographers and the Regional Science Association, Montreal, April 3, 1956

Derwent Whittlesey opened the discussion by distinguishing between two kinds of conceptual regions—the uniform regions, widely used by geographers, and the more recently defined nodal, or functional region. The uniform region is defined in such a way that its various characteristics are essentially homogeneous, whether they be natural characteristics or man-made. The nodal region is distinguished by a functional focus on a node (or series of nodes), with which is associated an area of activity related to the node by means of a complex of communication links, or ties. The node would normally be determined by the focusing of human activity, such as occurs at an urban center, and the limits of the region determined by the area of influence associated with the node such as a metropolitan area. On a larger scale than that of the city, Professor Whittlesey gave as an illustration the wheat growing area of the U.S. with the nodal focus at the wheat trading center in Chicago. On a smaller scale than that of the city, he suggested that a department store might constitute the nodal focus of a related market area. It is also possible that a nodal region may be defined in purely physical terms, as for example, a river system with the node at the mouth of the river. Professor Whittlesey concluded with the observation that a given area, such as the wheat growing area of the U.S., may be considered as a uniform region and as a nodal region.

Taking up from the general statement of the nodal concept, Walter Isard presented a synthesis of various theories of nodal relationships and regional definition. Professor Isard started with the traditional Weberian location analysis, using locational and weight triangles to determine the production point of minimum transport cost, to which he added the "polar" type of analysis developed originally by Palander. Into this combination were injected factors of additional commodities, additional production points, urbanization and agglomeration economies and diseconomies and the like. Professor Isard then superimposed onto the pattern thus far evolved, the concepts of spatial distribution of urban centers and economic activity as developed by Christaller and Lösch, and completed the synthesis with the addition of the concept of rings of agricultural activity as originated by von *Thünen*. In working his synthesis Dr. Isard emphasized that there is no fundamental inconsistency in the various theories of nodal structure.

The last speaker of the session, Howard L. Green, discussed the interaction concepts of nodal region. He suggested that an hierarchical structure of sites might be worked into the concept on the basis that various current theories (the central place concept, the theory of metropolitan regions, and gravity models) are three aspects of a single pattern of interrelationships between settlements. Dr. Green suggested arranging the centers of the U.S. in an hierarchy starting with the center of greatest activity where certain facilities are to be found there only (most likely New York, N.Y.) and dropping down by successive layers as various activities are duplicated in smaller centers. New York would appear on all levels of the hierarchy since its population is served not only with certain unique facilities, but also with the same kind of activities as are to be found in the smaller centers. Thus centers on each level would appear on all lower levels. Dr. Green observed that problems might arise in analyzing the interaction of two centers on a given level of the hierarchy if they are separated physically by a third center which also appears on the next higher level of the hierarchy. Dr. Green proposed empirical study using the number of long distance telephone calls between communities as a measure which would help define the existing pattern.

Gerald A. P. Carrothers, Reporter

Dear Regional Scientist: July 7, 1956, The Regional Science Association, MIT, Cambridge, Mass

Enclosed is a copy of our Constitution with various changes as suggested through correspondence and through discussion at last December's business meeting. The most important changes concern financial liability of members and the liberalization of requirements for membership. If there are any serious objections to the Constitution in its present form, I should be informed about them by August 15. According to the understanding reached at the business meeting, if the response to this draft of the Constitution by members of the Association as of December 28, 1955, indicates that no major changes should be effected, the Constitution will be assumed to have been ratified by the members, and we will proceed to appoint a nominating committee.

In connection with our *Papers and Proceedings*, we have been faced with a rather embarrassing problem since last fall. Many persons have written in, desiring to purchase a copy of Volume I, even at the price of $5.00 which is currently being charged. However, we have only a limited number of copies of Volume I, and these are being reserved for purchase by University libraries. We have had to decline other purchase requests.

In order to minimize this problem during the coming year, I would suggest that if you are likely to recommend anyone for membership in the Association, or if you are likely to suggest to anyone that he obtain a copy of Volume II of the *Papers and Proceedings*, you do so now, and let me have the names of these persons. In this way such persons will be placed on our mailing list before the printing order on Volume II of the *Papers and Proceedings* is placed, and thus will be able to obtain a copy of Volume II with a voluntary contribution of $1.00 …

The dollars are rolling in, and we are grateful for them. If you have not already contributed your dollar, let this serve as a reminder to do so.

Sincerely, Walter Isard

Constitution of the Regional Science Association

Article I. Name

The name of the organization shall be the Regional Science Association

Article II. Objectives and Scope

The Regional Science Association is an international association for the advancement of regional analysis and related spatial and areal studies. The

Association shall operate as an objective, scientific organization without political, social, financial, or nationalistic bias. Its main objectives shall be to foster exchange of ideas and promote studies focusing on the region and utilizing tools, methods, and theoretical frameworks specifically designed for regional analysis as well as concepts, procedures, and analytical techniques of the various social and other sciences. The Association shall support these objectives by promoting acquaintance and discussion among its members and with scholars in related fields, by stimulating research, by encouraging the publication of scholarly studies, and by performing services.

The Regional Science Association shall not participate in activities involving carrying on propaganda, or otherwise attempting, to influence legislation, nor shall it participate in, or intervene in (including the publishing or distributing of statements) any political campaign on behalf of any candidate for public office. No part of the various receipts of the Regional Science Association shall inure to the benefit of any private individual.

Article III. Membership

Section 1. Members
The principal criterion of eligibility for membership shall be a mature and deep interest in the field of regional science. Any person with such an interest and in agreement with the objectives of the Regional Science Association may be proposed for membership by transmittal to the Secretary of a membership application form, signed by two members of the Association. On receipt of the properly completed application form and payment of dues, the applicant's name shall be entered on the roster of members and the Secretary shall notify the applicant of this action.

Membership shall be renewed annually by payment of dues. If a member does not pay dues within six months after the date of official notice by the Treasurer, the membership shall be considered terminated and the name removed from the roster of members without further notice.

Section 2. Membership Rights
All members shall have full voting and office holding rights.

Section 3. Misuse of Affiliation
Any Member or Associate who makes use of his affiliation with the Association in a manner considered improper by the Council may be suspended by the Council after opportunity has been given the individual for a hearing before the Council and may be expelled from the Association by a vote of the Council and a majority of Members voting at the next Business Meeting of the Association.

Section 4. Limitation of Liability
No member or officer shall be individually liable for the debts, contracts, and other obligations of the Association, other than his dues paid for membership. This proviso shall appear in all contracts entered into by and on behalf of the Regional Science Association.

Article IV. Officers, Council, and Committees

Section 1. Elected Officers and Committees
The elected officers of the Association shall be a President, a President-elect, two Vice Presidents, a Secretary, a Treasurer, and six elected Councilors. The duties of these officers shall be those normally pertaining to their various posts. During the first year of this organization the President-elect shall serve also as President. A Nominating Committee shall be elected also. The terms of office shall begin on the day following the Annual Meeting of the Association and the period between Annual Meetings shall be considered a one-year term. The terms of office shall be one year for President, President-elect, Vice President, and members of the Nominating Committee, and three years for Secretary, Treasurer, and elected Councilors. The President, Vice President, and elected Councilors shall not be eligible for immediate reelection to the same office, and the Secretary and Treasurer shall not be eligible for reelection to the same office until after a lapse of six years following termination of the first tenure. The terms of office of the elected Councilors shall be arranged so that two shall retire each year.

Section 2. Methods of Nomination and Election of Officers
The Nomination Committee shall make two or more nominations for each office except for that of President-elect, for which office only one nomination shall be made. The Nominating Committee shall submit its slate of candidates to the Secretary at least four months prior to the next Annual Meeting of the Association. The Secretary shall immediately notify the membership of these nominations. Additional nominations may be made in writing by any twenty members of the Association if received by the Secretary at least 90 days prior to the announced date of the next Annual Meeting. At least 60 days before the Annual Meeting the Secretary shall mail to all Members official ballots to be returned to the Secretary within 30 days and be counted by tellers appointed by him from the list of Members. The Council shall have power to fill vacancies until the next election.

Section 3. Council and Executive Committee
The Council shall consist of the officers elected under Section 1, and the most recent past President. The Council shall have power to transact all business of the Association and to assign specific responsibilities to the

various officers and committees of the Association. The Council shall assist the President in selecting committees to advance the work of the Association and may designate certain committees to be elected by the membership. The Council may delegate to the officers authority to sign contracts. The Council shall meet at least once a year at the call of the President. Notices of Council meetings shall be sent out at least two weeks in advance. A majority of the Council shall constitute a quorum.

Within a maximum limit set by a two-thirds majority of the ballots cast by mail on a proposal circulated by the Secretary with the official ballot for election of officers the Council may assess and collect dues. The Treasurer shall submit, and the Council shall approve, a budget based upon a fiscal year from July 1 to June 30.

The Council shall elect an Executive Committee for the purpose of transacting the business of the Association. The President and President-elect shall be ex-officio members of the Executive Committee. Official actions of the Executive Committee shall be subject to the approval of the Council by majority vote in a mail ballot. Official actions of the Executive Committee and by the Council shall be published as promptly as practicable.

Section 4. Editors
An Editor or Editors shall be appointed by the Council for such term as it may designate.

Section 5. Committees
A Nominating Committee of three Members shall be elected at the Business Meeting of the Association. The Council shall make at least three nominations for members of the Nominating Committee. The nominations by the Council shall be announced to the membership the same time as the report of the outgoing Nominating Committee. For the Nominating Committee three additional candidates may be nominated by the membership at large. Such nominations may be made by mail if supported by at least ten Members, or at the Business Meeting, if seconded. The first three additional nominations shall be accepted.

Article V. Meetings

Section 1. Annual Meetings
The Annual Meeting of the Association shall be held at such time and place as the Council may designate. The Council may arrange other meetings in addition to the Annual Meeting. Announcement of the time and place of meetings must be mailed to Members at least 60 days in advance.

Section 2. Business Meetings
A Business Meeting shall be held during the Annual Meeting. During this Meeting, as part of the Secretary's report, a summary of the actions of the Council shall be presented to the membership. Reports shall be presented also by the President, Treasurer, Editors, and Chairmen of the active committees.

Article VI. Changes in the Constitution

Changes in the Constitution shall be proposed either by action of the Council or by petition of 25 members and shall be adopted upon affirmative vote of two-thirds of Members voting, where the vote may be taken in either of two ways: first, at any regular meeting by ballot mailed or handed to the Secretary, provided that printed notice of the proposed change was mailed to all Members with the call of the meeting; second, by mail ballot at any time, provided the ninety days' notice of the proposed change has been mailed to all Members.

The August 6 newsletter discusses the need to include relevant political variables in our regional analysis and accordingly announces once again a joint session with the American Political Science Association in Washington, D.C.; a brief paper on the *Division of Powers: An Areal Analysis* by Arthur Maass and initial comments by discussants are distributed to RSA members. The November 2 (not included below) and December 14 newsletters announce regional science fellowships at the University of Pennsylvania and provide information on the December Cleveland RSA program.

Dear Regional Scientists: August 6, 1956, Regional Science Association, MIT, Cambridge, Mass.

Over the last few years many of us have become increasingly aware that our techniques and models for regional analysis are limited in their usefulness because of our inability to embrace within them the relevant political variables. We have therefore felt the need for closer association with political scientists and for urging the incorporation of a spatial or regional framework into political theory and thinking, much as has been partly achieved over the last decade in economic theory and practice.

To further these objectives, the Regional Science Association will sponsor jointly with the American Political Science Association the following session to be held on September 6, 1956, 2:30 PM Hotel Statler, Washington, D.C. (the room of the meeting will be noted on the program of the American Political Science Association.)

Division of Powers: An Areal Analysis

Chairman: *Arthur Maass*, Harvard University

Papers:
1. A Reexamination of Certain Critical Writing on Division of Powers—Montesquieu, Tocqueville, and others
   Stanley Hoffman, Harvard University
2. Certain Criteria for a 'Proper' Areal Division of Government Powers
   Paul Ylvisaker, Ford Foundation
3. Application of Criteria for Areal Division of Government Powers to New York Metropolitan Region
   Robert Wood, Harvard University

Discussants: *Carl J. Friedrich*, Harvard University; *Walter Isard*, University of Pennsylvania; *Hugh Elsbree*, Library of Congress; *Arthur MacMahon*, Columbia University (invited); *Vincent Ostrom*, University of Oregon (invited)

In inviting participants and outlining the topics to be covered by each paper Arthur Maass has made some very rough notes which, however, present very lucidly and pointedly the materials and problems to be covered. I am therefore circulating them for your information. I hope you will be able to attend the session and participate in the discussion, which I am sure, Arthur Maass would welcome.

Sincerely, Walter Isard

Notes on Division of Powers: An Areal Analysis by Arthur Maass

Prepared June 1955 by Arthur Maass for the September 6, 1956 session

Theme: Since the time of Aristotle political science has been concerned centrally with the distribution and division of government powers. Today's leading text on constitutional government states at the outset that "Division of power is the basis of civilized government. It is what is meant by constitutionalism."

It has been traditional among political scientists to think of this problem principally in terms of the distribution of powers among the officials at the capital city of a defined political region—the separation of powers and checks and balances, for example; and little attention has been given to the distribution of government powers among levels of political and administrative regions. Yet, as Fesler has indicated, this areal division of power may be of equal importance to its division among legislature, executive, and other agencies in the central city.

Why have political scientists failed to produce theoretical and systematic analyses of the areal division of government powers? It may be because political science remains today very much influenced by the brilliant and broad system of analysis laid down by its greatest student. And, of course, there was little reason for Aristotle to be concerned with the areal distribution of power; the experience of the Greek *polis* cast little light or emphasis on this.

It may be also because of factors relating to the emergence of the nation state. The areal consolidations of power here were of course crucial, but difficult and in many cases tenuous. Thus, when division of states' powers—constitutionalism—was urged and justified by 17th and 18th century political scientists the emphasis was on a division among ruling bodies at the seat of government. It might have endangered the very continuance of the nation state to urge then a broad areal division. And in France and elsewhere a full realization of the rights of man appeared to necessitate an extinction of the areal distribution of powers which characterized the *ancient regime*. For these reasons, perhaps, Montesquieu, whose concern with separation of powers is classic in the literature of political science, gave little emphasis to its area manifestations.

It may be that the 20th century pluralist challenge to the doctrine of sovereignty provided, for the first time since the emergence of the modern nation state, a stimulus for study of the areal division of powers. But despite the writings of Laski and others, this stimulus has not been sufficiently sustained to yield systematic analytical results.

These speculations are interesting; but we are concerned with the development of an approach or framework which will lead to a theoretical and systematic analysis of the areal division of government powers, and we are interested in why political scientists have failed to do this in the past *only* insofar as this will help in the present task.

Thus, political scientists have concluded that separation or division of powers is essential for a constitutional democratic state. This conclusion is based on certain fundamental assumptions about modern government. Political scientists have applied these assumptions and the conclusion to dividing power at the seat of government. We seek now to apply them to the division of powers areally.

Regarding the first paper by *Stanley Hoffman* on a reexamination of certain of the classic writings on division of powers—Montesquieu Toqueville and others, the aim is to determine how the principle enunciated in these writings can be applied to current thinking and experience on the areal division of power.

Regarding the second paper *by Paul Ylvisaker,* the problem is how to divide power among levels in a hierarchy of regions without the need to identify the several levels with existing levels of government— i.e., town, county, state, nation.

Some possible broad categories of criteria:

1. Criteria that result from the requirement for constitutional government that government powers be divided—i.e., that result from the application to area of the assumptions that have led to divided power at a nation's capital. The requirements of popular control, checked and balanced powers, etc.
2. Criteria that result from the requirement of "citizenship" in the Aristotelian sense—i.e., popular participation in government.
3. Other criteria:
   a. Efficiency
   b. Authority needed to conduct a function—for example, Lilienthal's statement that TVA needed the full authority of the Federal Government to lick Commonwealth and Southern, but that the Administration of this authority could be decentralized.
   c. The extent to which physical environmental factors are determining—i.e., man's ability to transcend these.

The third paper by Robert Wood will consider the application of the theory and criteria developed to a concrete situation, namely the New York Metropolitan Region. This paper can be either an effort to develop an ideal division of powers for a hierarchy of governments in the Region, or an analysis and evaluation of the existing division in terms of the theory and criteria, or a combination of all or parts of each of these.

Dear Regional Scientists: December 14, 1956, Regional Science Association, Univ. of Pennsylvania, Philadelphia, PA.

Enclosed are announcements regarding the Regional Science program and Fellowships at the University of Pennsylvania. I would greatly appreciate your calling this program to the attention of promising graduate students. Inquiries should be addressed to me. Also, there follows an early statement of the Cleveland program. Note that instead of asking the question: what can each of the conventional disciplines contribute to regional science, we turn the question around and in the first session ask what can regional science contribute to each of four conventional fields.

Regional Science Association Program
Cleveland, December 27–29, 1956

THURSDAY, DECEMBER 27, 1956

9:00 A.M. Subject: *Potential Contributions of Regional Science*, Regional Science Association

Chairman: (to be announced)

Papers:

1. To the Field of Geography
   Thomas R. Smith, University of Kansas
2. To the Field of City Planning
   Louis B. Wetmore, University of Illinois
3. To the Field of Economics
   Joseph L. Fisher, Resources for the Future, Inc.
4. To the Fields of Sociology and Political Science: Some Survey Results
   Walter Isard, University of Pennsylvania

2:30 P.M. Subject: *Transportation Analysis for Metropolitan Regions,* American Economic Association and Regional Science Association

Chairman: *Paul N. Guthrie*, University of North Carolina

Papers:

1. Analysis with Particular Reference to Chicago and Detroit Regions
   J. Douglas Carroll, Jr., Chicago Area Transportation Study
2. Analysis with Particular Reference to the Philadelphia Region
   Robert B. Mitchell, University of Pennsylvania

Discussion: *William Goldner*, Bowling Green State University; *David L. Glickman*, Port of New York Authority; *Ernest Williams*, Columbia University

---

8:00 P.M. Subject: *Regional Topics: I*, Regional Science Association

Chairman: *George Ellis*, Federal Reserve Bank of Boston

Papers:

1. Long and Short Range Regional Forecasts
   Lawrence E. Fouraker, Pennsylvania State University
2. Projections for the Gulf Coast Region
   F. V. Walker, University of Southern California

3. Areal Functional Organization and Regional Geography
Allen K. Philbrick, Michigan State University

4. A Synthesis of Theories of Location, Transport Rates and Spatial Price Equilibrium
Earle W. Orr, Purdue University

FRIDAY, DECEMBER 28, 1956

9:00 A.M. Subject: *Interregional Flows of Funds*, Regional Science Association and American Statistical Association

Chairman: *Philip Neff*, University of California at Los Angeles

Paper by: *J. Dewey Daane*, Federal Reserve Bank of Richmond; *Norman Dowsher*, Federal Reserve Bank of St. Louis; and *Robert Einzig*, Federal Reserve Bank of San Francisco

Discussion: *Werner Hochwald, Washington* University; *Seymour E. Harris*, Harvard University; *Edward K. Smith*, Boston College

---

2:30 P.M. Subject: *Analysis for Underdeveloped Regions*, Regional Science Association

Chairman: *Harvey S. Perloff*, Resources for the Future, Inc.

Papers: 1. Population Growth and Economic Development in India
Edgar M. Hoover, Office of Population Research

2. Planning Theory and Regional Capital Budgets with Special Reference to Puerto Rico
Lloyd Rodwin, Massachusetts Institute of Technology

Discussion: *Bert F. Hoselitz*, University of Chicago; *Stefan Robock*, Midwest Research Institute; *Robert L. Allen*, University of Virginia

---

5:00 P.M. *Business Meeting*, Regional Science Association

8:00 P.M. Subject: *Regional Topics: II*, Regional Science Association

Chairman: *George T. Conklin*, Guardian Life Insurance Co.

Papers: 1. Location Factors in Industrial Development of Ports
Allan Rodgers, Pennsylvania State University

2. Location Theory and Empirical Evidence
Charles M. Tiebout, Northwestern University

3. Location Factors in the Development of Steel Centers
Paul G. Craig, Ohio State University

SATURDAY, DECEMBER 29, 1956

9:00 A.M. Subject: *Linear Systems in Regional Analysis*, Regional Science Association and Econometric Society

Chairman: *Harold J. Barnett*, Resources for the Future, Inc.

Papers: 1. Optimum Regional Agricultural Patterns
James Henderson, Harvard Economic Research Project

2. General Interregional Equilibrium
Walter Isard, University of Pennsylvania

Discussion: *Frederick T. Moore*, Rand Corporation; *Richard Nelson*, Oberlin College; *Robert E. Kuenne*, Princeton University

---

2:30 P.M. Subject: *Gravity Models*, Regional Science Association

Chairman: (to be announced)

Paper: 1. The Geography of Prices and Spatial Interaction
William Warntz, American Geographical Society

Discussion: *Alan M. Voorhees*, Automotive Safety Foundation; *Ernest H. Jurkat*, Marketers Research Service; *Reavis Cox*, University of Pennsylvania; *Gerald A.P. Carrothers*, University of Toronto

---

8:00 P.M. Subject: *Regional Topics: III*, Regional Science Association

Chairman: *Paul M. Reid*, Detroit Regional Planning Commission

Papers: 1. The Role of National, Regional, and Local Factors in Post-War Housing Markets
Reinhold P. Wolff, University of Miami

2. Regional Aspects of Economic Development in Multi-National States
Alexander Melamid, New School for Social Research

3. Influences on Regional Credit Expansion
Robert Emmer, Federal Reserve Bank of Chicago

5 The Rooting and Emergence of Regional Science as a Major Field of Study

The year 1956 saw three developments significant for the association and regional science in general. First, there was a major step forward in the *formal recognition of the association as a legitimate social science organization and consequently its incorporation in the Allied Social Science Associations group.* This involved the inclusion of its officers and December sessions in the combined ASSA printed program for the December 1956 meetings of nine associations. The nine were: American Economic Association, American Finance Association, American Marketing Association, Industrial Relations Research Association, American Farm Economics Association, Econometric Society, American Association of Teachers of Insurance, Regional Science Association—holding joint meetings with: Catholic Economic Association.

The second major development was the *establishment of a Ph.D. program in Regional Science at the University of Pennsylvania.* This was an outgrowth of the need to train scholars competent in the methods, approaches and models for regional research and urban and regional policy analysis. By 1956 it was widely recognized, as indicated in the Bessey letter above and in the reply, that effective research requires knowledge of useful methods and approaches that come from several disciplines; these methods, as a set, were not being taught in any existing Ph.D. program. The need for multi-disciplinary training of students was crystal clear. At Harvard, graduate students had the opportunity to dig deeply into industrial location and regional economic development. However, while they took courses with Professor Joseph J. Schumpeter, Wassily Leontief and other distinguished faculty in meeting the requirements of a Ph.D program in economics, they had very little opportunity to obtain exposure to geography, sociology, political science and planning analysis. They, Kuenne, Kavesh, Lindsay, Airov and others were trained strictly as economists. At MIT where location theory and urban analysis was taught, students with a planning interest who wished to undertake a Ph.D. program (such as Coughlin, Stevens and Carrothers) had to enroll in the Economics Department. While there they had exposure to the outstanding teaching of Professors Paul A. Samuelson, Robert Solow and others in the relatively new mathematical economics, they had even less opportunity to have exposure to geography, sociology and political science.

Thus, at MIT Professors Lloyd Rodwin and Isard proposed to establish within the City Planning Department of the School of Architecture and Planning a Ph.D.

program in Planning Analysis with a multi-disciplinary focus, at least one with a more balanced social science framework than existing Ph.D. programs. The proposal encountered rough sledding owing to not only the usual obstacles to establishing a new Ph.D. program at a premier institution, but also internal difficulties within the Planning department which in its search for a new chairman chose an Australian scholar whose appointment became entangled with U.S. immigration policies. Hence, after a set of discouraging delays at MIT, the opportunity to establish a Ph.D. program at the University of Pennsylvania was seized upon by Isard with enthusiasm, and the offices of the Regional Science Association were accordingly relocated. The advent of regional science and the Association helped bring new life to the University of Pennsylvania's long declining economics department and its Wharton School, in which the regional science program was located.[12]

An early statement of the regional science program at the University of Pennsylvania follows:

Regional Science, Walter Isard, Chair

The graduate program leading to the degree of Doctor of Philosophy in Regional Science is administered by an interdisciplinary committee composed of: Dean Roy F. Nichols; Morris Hamburg, Economic Statistics; Walter Isard, Economics; Lester E. Klimm, Economic Geography; Irving Kravis, Economics; E. Frank Stover, Civil Engineering; Stephen B. Sweeney, Political Science; and William L.C. Wheaton, City Planning.

The degree of Doctor of Philosophy in Regional Science signifies the attainment of a high level of scholarship and ability to conduct independent research in this field. The basic program centers around coursework in location, space and regional theory, active participation in the Regional Science Seminar, and research activity within the regional science group. Each student should arrange his program in consultation with the chairman of the interdepartmental committee. This program should consist of courses designed to prepare the student adequately for the preliminary examination and of training to develop his ability to undertake and complete independent research with distinction, to be demonstrated in the doctoral dissertation.

[12] In Novosibirsk and its Akademgorodok, the foremost academic center in the Soviet Union, the Wharton School of Finance and Commerce later became known as the Wharton School of Regional Science.

Requirements for the Degree of Doctor of Philosophy

Prerequisites for Admission

In addition to the general requirements for admission to the Graduate School, the student undertaking study for the Doctor of Philosophy in Regional Science must have completed, on the undergraduate level, one year of calculus or its equivalent, one year of statistics, and one year of intermediate theory in a social science field. Exceptional applicants may be admitted for graduate study pending fulfillment of these entrance requirements.

Preparation for Preliminary Examination

1. Examination must be passed in two modern foreign languages in accordance with Graduate School regulations. Language requirements should be met as early as possible in the program. Special attention is called to the time requirements of the Graduate School.
2. A minimum of 54 semester credits (including thesis) which must be earned in courses approved by the Graduate School of Arts and Sciences. Of these semester credits at least 16 must cover graduate work in the theory of regions and regional science partly utilizing courses in economic theory and city planning; at least 8 must cover other courses in statistics and quantitative method; and at least 12 must cover other courses in one of the following fields: political science, sociology, geography, and city planning. At least 30 of the 54 semester credits must be completed in the Graduate School of the University of Pennsylvania, and 24 of these must be completed before the preliminary examination.

The Preliminary Examination

In addition to the other requirements previously stated, admission to candidacy will be achieved by passing a preliminary examination. The examination will consist of three parts, of which the first will have double weight:

A. Theory and method in regional science including relevant elements of theory in related social science fields.

B. Statistics, theory of, and application to regional problems.

C. An applied field such as Transportation, Land Use, Industrial Location, and Population.

Dissertation and Final Examination

Students who are accepted as candidates will complete an approved thesis, demonstrating a high level of ability to undertake and complete original research, on a topic in one of the fields of the preliminary examination. The facilities of the Institute for Urban Studies, the Wharton School and other branches of the University are available for doctoral research work.

A final oral examination will be passed on the subject of the thesis and closely related fields.

Courses

The courses listed below are those which are either required of, or recommended for all degree candidates. Additional descriptions of the courses and prerequisites will be found under the offerings of the respective departments.

Regional Science

701-2 Seminar in Regional Science. Fall and Spring terms. 4 s.c. Isard and staff. Theories of regional structure, research methods and empirical analyses of regional problems.

City Planning

522 City Planning Organization and Effectuation. Spring term. 2 s.c. Mitchell

544 Development and Structure of the Contemporary Community. Fall term. 2 s.c. Rapkin

551 Housing. Fall term. 2 s.c. Wheaton

557 Urban Land Utilization. Spring term. 2 s.c. Rapkin

628 Theory of City Planning. Fall term. 2 s.c. Dyckman

672 Urban Transportation. Spring term. 2 s.c. Mitchell

Economic Theory

603 International Economics. Both terms. 4 s.c. Kravis

604 Recent Developments of Economic Theory. Both terms. 4 s.c. Weintraub

609 Economic Development. Both terms. 4 s.c. Easterlin

610 Inductive Economics. Both terms. 4 s.c. Klein

611 Introduction to Mathematical Economics. Both terms. 4 s.c. Klein

618 Industrial Location and Regional Development. Both terms. 4 s.c. Isard

Economic Statistics

659	Seminar in Economic Statistics. Both terms. 4 s.c. Friend
663	Sample Survey Methods. Both terms. 4 s.c. Tepping
664	Intermediate Statistics. Both terms. 4 s.c.
667	Sequential Analysis and Nonparametric Tests. Spring term. 2 s.c. Clelland
668	Advanced Sampling Theory, Design of Experiments, and Special Topics in Statistics. Both terms. 4 s.c.
669	Quantitative Methods of Business Planning. Both terms. 4 s.c. Tepping

Economic Geography

931	Geography of Industrialization. Both terms. 4 s.c. Klimm
933	Economic Geography—World Pattern of Economics. Both terms. 4 s.c. Tepping

Political Science

622	Theory of Public Administration. Both terms. 4 s.c. Sweeney, Davy, Bent
686	Research Methods in Political Science and Public Administration. Fall or Spring term. 2 s.c. Carter
784	Seminar in Political Theories. Both terms. 6.s.c.

Sociology

601-2	Contemporary Sociological Theory. Both terms. 4 s.c. Bressler
615	Population. Both terms. 4 s.c. Hutchinson
780	Population Seminar. Both terms. 4 s.c. Thomas

In this program, the focal point was the Regional Science Seminar. There, relevant methods of the several social sciences, engineering fields (civil and transportation) and operations research were discussed, providing the integrating thrust of regional science.

The third major development in 1956 was the establishment of the *Regional Science Research Institute.*

From experiences at Harvard University and MIT, the lesson was learned that it could be extremely difficult to obtain approval of a new initiative at top-notch universities, in particular in proposing and administrating new research projects and in publishing new journals and other outlets for scholarly publications. This led Isard with the assistance of Benjamin H. Stevens to establish in 1956 the Regional Science Research Institute, a non-profit organization outside, but physically nearby, the University of Pennsylvania. The major purpose for which the Institute was founded was to conduct "multidisciplinary studies of the spatial and locational interaction and interdependence of economic, social, political and environmental phenomena associated with urban development and regional growth."

Specific objectives of the *Institute* were to "advance through the following major areas of endeavor:

1. Abstract and conceptual research on theories and models of location and spatial interaction
2. Applied research on, and studies of, specific problems and geographic areas.
3. Research training of graduate and post-doctoral students through their employment on research projects and participation in Institute seminars and by providing direction of doctoral dissertations and other research papers.
4. Dissemination of theoretical and applied research results through the publication of the *Journal of Regional Science*, monographs, bibliographies, reports, and discussion papers
5. Informal and uncompensated advice to government and academic institutions on policy and research issues."

The Institute, largely administered by the socially engaging and stimulating Benjamin H. Stevens contributed in a major way to the research training of regional scientists who subsequently played (and still play today) a significant role in the development of the Regional Science Association.

The publication of the first regional science journal by the Regional Science Research Institute, which later became the premier journal in regional science under the outstanding guidance by Ronald Miller, the managing editor, began in 1958. The formal announcement of it was:

Journal of Regional Science

As many of you are aware, there have been presented at sessions held jointly with diverse associations a number of fine papers which have not been published. (The *Papers and Proceedings* contains only those papers presented at the annual Christmas meetings). There have been many requests for these papers. Moreover, there has come to my attention a number of first-rate manuscripts in the field of regional science, which because of the methodology employed, subject considered, or both, have not been published in existing journals; or have not been published in a satisfactory manner. Because of this undesirable situation, those of us associated with the Regional Science study program at the University of Pennsylvania have decided to publish a *Journal of Regional Science*. This journal will be issued as frequently as top-quality manuscripts are available, probably three or four times a year. We hope to avoid the publication of mediocre manuscripts just in order to maintain a regular publication schedule—a lot which too frequently befalls the best of social science journals. William Alonso and I will serve as editors of the Journal. The contents of the first issue of the Journal, to be published this Spring, have already been selected.

Those who wish to submit manuscripts to be considered for publication in subsequent issues should send those manuscripts to me at the Wharton School, University of Pennsylvania, Philadelphia 4, Penna. Topics to be covered by the Journal will be in the same broad range as those covered in the *Papers and Proceedings of the Regional Science Association*; outstanding scholarly manuscripts relating to new aspects of regional structure and function will be most welcome. No limit will be imposed on the length of papers, but the expected range is from 4,000 to 15,000 words. Longer articles will be published on the basis of unusual excellence. The author must supply proofread, typewritten copy of the text, and be responsible for the provision of any necessary figures and tables suitable for direct photographic reproduction. In general, we will reach a prompt decision on manuscripts submitted. Consideration of manuscripts will start April, 1958.

Formal responsibility for the publication of the *Journal of Regional Science* will be undertaken by the Regional Science Research Institute, a non-profit corporation incorporated under the laws of the State of Delaware on September 24, 1956. All members of the Regional Science Association who remit for year 1958 $2 as annual dues and payments will be entitled to a free copy of each 1958 issue of the Journal, one-half of this $2 sum being payment for the Journal by vote at the business meeting of the Regional Science Association.

Libraries, businesses, and other non-members (only individuals are eligible for membership in the Association) may order subscriptions to the Journal from: Regional Science Research Institute, Benjamin H. Stevens, Treasurer, GPO box 8776, Philadelphia 1, Penna. Invoices will be rendered after the first copy has been sent. The Institute will be happy to accept continuation orders under which subscriptions will be automatically renewed each year.

Invoices will then be rendered after the first issue of each year. Price: $5.00 per year (minimum 2 issues), $2.50 single issue.

Please note that non-members must place two separate orders if they wish to receive both the *Journal of Regional Science* from the Regional Science Research Institute and the *Papers and Proceedings* from the Regional Science Association.

Walter Isard

For the period, 1965–1969, among economic and related journals, this flagship journal was ranked fifth for citations in articles, closely behind Harvard's *Review of Economics and Statistics* and its *Quarterly Journal of Economics*.

Another major force contributing to the development of the Regional Science Association stemmed from the 5-year *Resources for the Future research grant* received by Isard. The work, which began in 1955 at MIT, was later shifted to the University of Pennsylvania. The proposed research to be conducted under this grant was as follows.

Research Proposal on Principles and Methods of Analysis for Resource Problems with Particular Emphasis on Regional Resource Problems

A. Significance of this Project

It is generally recognized among students of resource problems that the uses of our diverse human and natural resources are complexly interrelated. It is also generally recognized that wiser and more efficient resource policies must go hand in hand with increased understanding of these interrelations, with increased knowledge of their quantitative importance. This project is designed to: (1) increase our understanding of these interrelations; (2) make possible sounder estimates, even if only rough, of the relative magnitude of the various factors associated with any current or potential resource problem; and (3) permit the clearer perception of the vital elements in programs aimed at balanced regional development and optimum resource utilization.

(For a specific illustration refer to the cursory discussion of the water problem of the West in Supplementary Statement A, pp. 5–10.)

B. Specific Aims

1. to study systematically and comprehensively existing and emerging concepts, methods, tools, techniques of analysis and theories in the several social sciences, particularly economics, in order to develop more productive approaches and methods in acquiring insights in the solution of pressing resource problems. The problem around which the proposed research would chiefly center are those associated with the larger-type geographic regions. The approaches and methods to be developed, however, would be directly applicable to the analysis of problems of urban-metropolitan regions as well, and would have major uses in problem studies concerned with:
   a. the improvement of principles and methods for evaluating resource development projects and programs, both public and private (R.F.F. problem area #2)
   b. The correction or mitigation of unemployment and distress in certain areas or industries heavily dependent upon waning natural resources (R.F.F. problem area #3)
   c. the role of energy in the United States economy (R.F.F. problem area #6)
   d. the improvement of the administration and management of resource enterprises and programs (R.F.F. problem area #7)
   e. the possibilities for economic and social development in smaller watersheds (R.F.F. problem area #9): and, among others,
   f. the impact of national tariff policy upon specific industries and local areas, and thereby upon the national economy as a whole.
2. To furnish improved analytic frameworks upon which resource studies could draw in their empirical research and to whose further improvement resource studies could contribute by furnishing additional testing materials and suggestions of superior hypotheses.

C. Method of Procedure

It is proposed that at M.I.T. and at the Washington offices of Resources for the Future the following techniques, theories, methods and approaches be carefully studied to identify their stronger elements and the potential usefulness of these elements for resource problem studies:

1. Gross Regional Product (GRP) projection techniques
2. Regional and inter-regional input-output (inter-industry) analysis; requirements approach.
3. Linear programming (activity analysis): operations research techniques; game theory
4. Spatial interaction and gravity models
5. Industrial Complex analysis
6. National Bureau of Economic Research methods; average and marginal (historical trend) ratio projections; demographic techniques.

(These broad investigations would extend and go beyond the traditional forms of analysis in industrial economics, regional and urban economics, location theory, economic geography and other fields.)

As the stronger elements of each technique, theory, method or approach are identified, ways in which they may be fused and interwoven into an eclectic but superior approach would be explored. The step-by-step construction of superior approaches via the synthesis of these elements is illustrated in the Appendix of this proposal.

Joseph L. Fisher, who was a very active and valuable member of the working committee on the organization of the Regional Science Association, and who was first associated with the Council of Economic Advisers (U.S. Government) and later with Resources for the Future, was extremely supportive of the proposal. The research conducted under the grant came to be largely embodied in the widely distributed book *Methods of Regional Analysis: An Introduction to Regional Science* (1960). As noted earlier, the MIT Press undertook the publication of this book. Its success led the Press to establish a *Regional Science Studies Series* which served for some years as an outlet for publication of quality works by regional scientists.

The Association, the Regional Science Research Institute and Department of Regional Science continued to grow steadily through the years 1957, 1958 and 1959. The newsletters of 1957 through 1959 and the annual December programs depict this. The newsletters are not reproduced below; only selected items from them are noted. Also, because the annual December programs have assumed a standard form, only that program for 1957 will be presented below in the regular text. Programs in the United States for 1958 and the following years (except 1963) will be presented in Appendix E.

Program: Regional Science Association
Hotel Sheraton, Philadelphia, Pennsylvania December 28–30, 1957

SATURDAY, DECEMBER 28, 1957

9:00 A.M. Subject: *Interregional Equilibrium and Linear Programming*, Joint Session: Econometric Society and Regional Science Association

Chairman: *Werner Hochwald*, Washington University (St. Louis)

Papers:
1. The General Equilibrium of Production, Transportation, and the Location of Industry
Louis Lefeber, Harvard University
2. Interregional Linear Programming
Benjamin H. Stevens, University of Pennsylvania
3. Aspects of General Interregional Equilibrium
Walter Isard and David Ostroff, University of Pennsylvania

Discussants: *James M. Henderson*, Harvard University; *Charles Tiebout*, Northwestern University

SATURDAY AFTERNOON, DECEMBER 28, 1957

2:30 P.M. Subject: *The Core and Boundaries of Regional Science*

Chairman: *Morris Garnsey*, University of Colorado

Paper: by the Regional Science Seminar, University of Pennsylvania (*W. Alonso, D. Bramhall, W. Isard, and B. Stevens*)

Discussants: *Lloyd Rodwin*, Massachusetts Institute of Technology; *Lyle E. Craine*, University of Michigan; *Preston E. James*, Syracuse University; *Maynard Hufschmidt*, Harvard University

SATURDAY EVENING, DECEMBER 28, 1957

8:00 P.M. Subject: *Regional Economic Studies*

Chairman: *David C. Melnicoff*, The Pennsylvania Railroad Co.

Papers:
1. An Approach to Measuring Regional Growth Differentials
George H. Borts, Brown University

2. Long-Term Regional Income Changes: Some Suggested Factors
Richard A. Easterlin, University of Pennsylvania

3. A Theory of Regional Social Accounting
Charles L. Leven, Iowa State College

SUNDAY MORNING, DECEMBER 29, 1957

9:30 A.M. Subject: *Methodological Issues in Regional Science*

Chairman: *Lester E. Klimm*, University of Pennsylvania

Papers: 1. Regional Differentiation and Socio-Economic Change
Otis Dudley Duncan, University of Chicago and *Ray P. Cuzzort*, University of Washington

2. Regional Development and the Geography of Concentration
Edward L. Ullman, University of Washington

Discussants: *Morgan Thomas*, Montana State University; *Richard Pfister*, Massachusetts Institute of Technology

---

SUNDAY AFTERNOON, DECEMBER 29, 1957

2:30 P.M. Subject: *Local and Regional Impact Studies*

Chairman: *Joseph L. Fisher*, Resources for the Future, Inc.

Papers: 1. Dependence of Local Economies Upon Foreign Trade
Werner Hochwald, Washington University

2. Resource Development: Regional Incidence of Costs and Gains
John V. Krutilla, Resources of the Future, Inc.

Discussants: *John Nixon*, Department of Commerce, State of New York; *Paul McGann*, U.S. Department of Interior; *Robert A. Kavesh*, Chase Manhattan Bank

---

5:00 P.M. *Business Meeting,* Regional Science Association

SUNDAY EVENING, DECEMBER 29, 1957

8:00 P.M. Subject: *Selected Regional Topics*

Chairman: *John C. Honey*, Carnegie Corporation

Papers: 1. The Government Sector in Metropolitan Projections
Robert C. Wood, Massachusetts Institute of Technology

	2. Recent Development in Central Place Theory *Brian J. L. Berry and William Garrison*, University of Washington
	3. Industrial Structure of American Cities *Irving Morrissett*, Purdue University
	4. Efficient Transportation and Industrial Location *Thomas A. Goldman*, RAND Corporation

MONDAY MORNING, DECEMBER 30, 1957

9:00 A.M.	Subject: *Regional Planning and Development*
Chairman:	*William L. C. Wheaton*, University of Pennsylvania
Papers:	1. Some Thoughts on the Planning of Metropolitan Regions *Gordon Stephenson*, University of Toronto
	2. Limits of Rational Action in Regional Development *Martin Meyerson*, Harvard University
Discussants:	*Werner Hirsch*, Washington University; *Irving Fox*, Resources for the Future, Inc.; *Edmund N. Bacon*, Philadelphia City Planning Commission

MONDAY AFTERNOON, DECEMBER 30, 1957

2:30 P.M.	Subject: *Gravity Models*
Chairman:	*Arthur T. Row*, Philadelphia City Planning Commission
Paper:	1. Population Projection via Income Potential Models *Gerald A.P. Carrothers*, University of Toronto
Discussants:	*Howard Bevis*, Chicago Area Transportation Study; *John Q. Stewart*, Princeton University; *Ralph W. Pfouts*, University of North Carolina; *Benjamin Chinitz*, New York Metropolitan Region Study.

One problem arising from growth was that of finances. Recall that at the business meeting on December 28, 1955 a motion was passed to request each member of the new association to contribute one dollar to meet the Association's expenses. Our editor, Gerald A. P. Carrothers, then produced in printed form Volume II of the Papers and Proceedings, where again each member of the association received a free copy. To libraries and non-members, sales of this volume were at $5.00 per copy, while sales of Volume I, because of large demand, continued to be restricted to libraries at $5.00 per copy.

In 1957 annual dues payment of $1.00 was formalized with a provision that only members who paid their dues by the date of publication of Volume III (dated 1957) covering the December 1956 meeting (whose program has already been presented) would receive a free copy of Volume IV of the Papers and Proceedings.

By 1958 the Journal of Regional Science was being published. As a consequence the annual dues and payments was raised to $2.00 per member. Of this amount, $1.00 was for membership in the Regional Science Association which entitled each member to a free copy of Volume IV of the Papers and Proceedings, and $1.00 for a subscription to the 1958 issue of the Journal of Regional Science. Only payments of at least $2.00 were taken to be acceptable since separate payments of $1.00 would double the work of the unpaid treasurer and secretary. Also, to reduce their workload, members were requested to have their libraries place standing orders for the Papers and Proceedings. At the December 1958 business meeting, the annual dues (covering subscription to the Journal of Regional Science) was raised to $4.00 per member when paid by September 1, 1959, or $5.00 after that date.

During the 1957–59 period there were held, in addition to the regular December meetings, a number of joint sessions. In 1957 they were with the Association of American Geographers (on *Nodal Regions*), the Eastern Sociological Association (on *Structure of the Metropolitan Region*), the American Institute of Planners (on *Economic Base Techniques*), and the American Political Science Association (on the *Development of American Federalism: A Forward Look*). The last joint session was a response to an awareness that continued to mount that our techniques and models for regional analysis were limited in their usefulness because of our inability to embrace within them the relevant political variables. We therefore felt the need for closer association with political scientists and for urging the incorporation of a spatial or regional framework into political theory and thinking, much as we had partly achieved over the previous decade in economic theory and practice.

In 1958 joint sessions were held with the American Institute of Planners where lectures for educational purposes were given on 1) Gravity, Potential and Spatial Interaction Models and 2) Interregional Linear Programming. The objectives of these lectures were to prepare members to understand better the more advanced papers to be presented at the December 1958 meetings.

One of the consequences of these joint sessions with planners was the preparation of the following statement.

Needed Metropolitan and Urban Research

A group of members of the Regional Science Association and others interested in Metropolitan and Urban Research have felt for some time that the exchange of information, the research direction, as well as the treatment of the subject matter at meetings could be considerably improved for the benefit of scholars in the field and of those dealing with concrete problems. As this research area is clearly best approached on an interdisciplinary basis, it was thought that a broad-gauged kind of organization should undertake to fill those gaps that most need to be filled. The Regional Science Association is such an organization, but ways in which it can better serve the field of metropolitan and urban research should be worked out.

If this were accomplished, it would tie in with the type of question asked repeatedly, and most poignantly again by Lloyd Rodwin at the Chicago meetings, as to where the Regional Science Association is going. It could well be argued that, at this stage of its development, the Regional Science Association has reached a point of achievement in the direction of devising certain general approaches, which would make it advisable to devote a portion of its activities to some few but specific interdisciplinary areas, of which metropolitan research could easily be chosen as the first. Such areas should be characterized by welding together the contributions of diverse disciplines to solve a concrete problem of operational significance which appears to be in need of solution before decisions can be made.

Nobody is suggesting that the Regional Science Association discontinue the presentation of high-level papers—in fact, the group felt that the discussions at last year's business meeting might have been responsible for papers at two sessions now concentrating on some metropolitan problems—but that in this field it might add another dimension. Nor is it suggested that in this area the Regional Science Association descend to the level of serving everyday operational needs in a semi-vocational way. What we need to develop primarily is the shaping of tools and methodology. But this should be put to use (e.g., the use to which Werner Hirsch's findings, presented at the ASA meetings in Chicago, can be put). Therefore, operational and action needs for research answers should be explored and evaluated before at least some research decisions are made. And then it is hoped that the Regional Science Association would concentrate on those which can best benefit by the contribution of several disciplines. Eventually progress in joint research methodology may emerge.

Thus, if research answers for broad operational needs can be found, then the Regional Science Association will have moved in this field in the direction of contributing to the needs of decision-makers in metropolitan areas as well as to the development of concrete interdisciplinary research programs—a contribution over and above RSA's present contribution to the individual fields represented in it.

It was therefore decided by the group to suggest to the Regional Science Association the appointment of a committee which would explore over several months first, what exchange of information, stimulation of research and types of meetings would best serve the specific metropolitan field, and secondly how such gaps could be filled by the Regional Science Association and in what manner. The business meeting unanimously approved the appointment of such a committee, with the understanding that whatever was or could be performed by other organizations should under no circumstances be duplicated but at best coordinated by the Regional Science Association. It was also suggested that, if successful, this or a similar committee might later on look into other areas of research needs, or become a research advisory committee for the Regional Science Association.

At this time the needs felt by the group can only be stated in general terms, prior to canvassing the opinions of a much larger number of people in the field, which the committee intends to do.

There is, first, a greater need for closer contact between those in need of concrete but imaginative answers to broad operating problems, and those planning all manner of research programs (from a long-term foundation support—to a summer project). Such contacts could define areas for some action-oriented research (e.g., on tax implications of rehabilitation), the areas where our definition of problems and our research methodology is weakest, and possibly come to some understanding of what kind of many needed investigations should take priority.

It is not being suggested to compile an exhaustive list of all needed research, but we might attempt to ask what questions are most in need of being answered tentatively now in order to be of use in broad decisions which must be made on, e.g., problems of long-term development of an area. The stimulation of research projects may, of course, be the result of this activity if successful.

The second area is one of mutual information. Not only a better clearinghouse is needed for big research activities going on anywhere in the field, but a much more detailed knowledge of the many small, incidental pieces of investigation carried out by consultants or individual researchers, not sufficiently broad to be listed as "research project under way", but which might provide many partial answers to burning operational problems (e.g., the benefits of residential versus industrial redevelopment). This type of service is only partly performed by existing organizations. It will take some imagination to find a way to fill that need with the necessary economy of volunteer time and without involving more than nominal expenses.

Reference has been made to the interdisciplinary nature of the metropolitan field. This should not be understood as an invitation to long discussions about each discipline's potential contributions. What is hoped for is that concrete problems can be defined and that the economists, political scientists, geographers, planners, etc., will be able to state both the research contribution which can (and should) be made by their own specialty, and to which questions they expect an answer from which other discipline. The best interdisciplinary work can only be done when each specialty has explored its own research potentialities on a particular problem. (If, e.g., the desirable future development of growing suburbs is to be fruitfully discussed, we need to know from the economist the financial structure of the whole area and its economic base, in comparison with other metropolitan areas; we should know the population projection from the demographer, the possibilities of re-structuring jurisdictional relations from the political scientist, the cost of benefit allocation between different alternatives from the resource specialist, and the potentialities of planning and zoning decisions from the planner.)

This leads into the third area where more effective work might be done eventually: if one big problem were selected for exploration for the better part of a year by several specialists, working individually, each attacking it with the tools of his own discipline, but meeting several times for exchange of ideas, mutual criticism and joint focusing on the main problems, then the presentation of their joint findings at the annual meetings would constitute a *new* contribution to research by the Regional Science Association, in addition to the individual papers now presented. If such a meeting (and possibly others in this field) were conducted without long-winded reading of papers, but thorough brief resumes of previously distributed papers followed by a thorough-going exchange of ideas by those having received and read them in advance, some real progress might be made. Also, devoting one of the meetings to free exchange of ideas on a specific problem might actually help to bring operators and researchers together in this new field. This, in turn, could over time lead to much-needed joint planning of research designs.

This is of necessity a very general statement of *some* of the needs which the group felt existed. It was designed purely to point the way, and it is hoped that specifics as well as additional areas and gaps which need to be filled will emerge as those questions are being explored.

The committee expects to solicit general ideas as well as concrete research proposals from a large group of workers in the vineyard of metropolitan research, and then hopes to be able to work them into concrete proposals for action by the Regional Science Association. Towards this end, answers to the following questions would be most welcome.

1. What are your reactions to the three areas of potential contributions outlined above.
2. What other gaps are you aware of which you feel can be filled by professional groups.
3. To what extent do other organizations fill some of these needs and how can their work be coordinated to avoid duplication?
4. Given your answers to questions 1 and 2, how would you suggest they can best be implemented by an organization concerned with metropolitan research, and specifically by the Regional Science Association? Concrete suggestions will be greatly appreciated.
5. What are the areas which you consider most important on which our present state of knowledge indicates that research needs to be done? This includes research needed for the sake of solving concrete action and decision problems which you are aware of; secondly, what research seems most urgent in terms of methodology and scientific advances in this field? Please name these areas in order of your own set of priorities. Would you also indicate in what way your own specialty can contribute to it, and what contributions you would expect from other disciplines?
6. What research (small or large) are you yourself connected with in a direct or indirect way or what on-going projects are you aware of which would not normally be picked up by the Urban Research Digest? How and through whom can more information be obtained?
7. Please let us have the benefit of any ideas you may have as to ways in which this committee may best go about its difficult task of canvassing opinions and information and welding them into workable proposals. In particular please name *all* those persons who in your opinion should be contacted because of their knowledge or experience in this field as well as their judgment.

The above statement was presented in discussion at the December 1958 business meeting of the Regional Science Association. As a result of the exchange of views that ensued a motion was passed unanimously to appoint, subject to confirmation by the Council, a sub-committee for the coming year for the purpose of investigating ways in which the Regional Science Association could better serve the area of metropolitan and urban research. This motion was passed on to the Council, which was provided with the above statement on the need for metropolitan and urban research and recommendation for the formation of a sub-committee. Also the Council was informed that the *group* of interested scholars who had prepared the statement had canvassed some of their members and others as to their willingness to serve on the sub-committee. The group suggested the following scholars from a variety of different disciplines to serve on such a sub-committee (later designated a committee), all of whom were willing to do so. To this list Benjamin H. Stevens was added to represent the field of Regional Science.

- Jesse Burkhead (Finance and Economics), Syracuse University
- Philip Hauser (Sociology), University of Chicago
- Werner Hirsch (Economics), Resources for the Future and Washington University
- Harold Meyer (Planning), University of Chicago
- Jerry Pickard (Geography), Washington Board of Trade
- Robert Ryan (Business and Economics), Committee for Economic Development
- Benjamin H. Stevens (Regional Science), University of Pennsylvania
- Robert Wood (Political Science), Massachusetts Institute of Technology
- Kirk R. Petshek, Chairman (Economics and Public Administration), City of Philadelphia

On April 30, 1959 the Council approved the appointment of this committee, with the understanding that such a committee shall operate without financial support from the Association.

In 1959 still another joint session was held with the American Institute of Planners. There the interest in research on urban and metropolitan problems continued to be intense.

6 The Invasion of and Extensive Expansion in Europe Concomitant with the Formation of Sections

During the late 1950s interest in the Regional Science Association, the Regional Science Research Institute and the Department of Regional Science and their programs also mounted rapidly in the international community. Along with the Bellagio conference about which more will be said later, there followed in the summer of 1960 lectures and papers given by regional scientists at several scholarly meetings in Europe. These meetings took place at the Hague, Paris, Bellagio Zagreb, Warsaw, Stockholm and Lund. The newsletters of March 21, Summer 1960 and October 20, 1960 report relevant details.

Regional Science Association, March 21, 1960 Wharton School, University of Pennsylvania, Philadelphia 4, PA.

Dear Regional Scientist,

This newsletter is to inform you of various meetings of the Regional Science Association to be held during the current year.

During the summer we will be holding a number of meetings (some small, others fairly large) in European cities. The first meeting will take place at the Hague, probably on June 7 or 8. A second meeting is scheduled in Paris at the Institute of Applied Economics, probably on June 14 or 15. A third meeting is set for July 2 at Bellagio near Milan, Italy following an EPA Conference on Regional Economics and Planning. A fourth meeting will probably be arranged at Warsaw during late July or early August. A fifth meeting is to be held in Stockholm, where we have already arranged for a joint session with the International Geographical Union when it convenes during the week of August 8 to 13. A sixth will be probably be scheduled for Lund, Sweden perhaps jointly with the Urban Geography Symposium, sometime during the week of August 15 to 19.

Each of these meetings is designed to promote interaction and exchange of ideas. Most participants are likely to come from nearby areas though it is hoped that there will be some from the United States and other distant countries, who may be able to attend and participate by presenting papers or taking part in the discussion. I shall be present at each one of these meetings and hope that at each meeting other regional scientists will be willing to present papers. Any persons who do wish to participate should contact Professor Duane F. Marble, Department of Regional Science, University of Pennsylvania. Professor Marble will be program co-coordinator for these various sessions. I believe that these meetings can be extremely fruitful in generating new ideas and urge you to attend.

On Tuesday evening, October 25, the Regional Science Association will have its annual joint session with the American Institute of Planners at 8:00 p.m. in Philadelphia. Details of this meeting will be forthcoming. Our own annual meetings are scheduled for St. Louis, December 27-30 when the American Economic Association and other allied social science organizations will convene.

Walter Isard

Regional Science Association, Summer 1960, Wharton School, University of Pennsylvania, Philadelphia 4, Pennsylvania

European Meetings of the RSA:

I. The Hague

Date: June 7 (9–12 AM, 2–5 PM)

Place: The Institute of Social Studies

Program: "Emerging Models in Regional Science", *Walter Isard*, University of Pennsylvania, U.S.A.

"A Few Criteria for the Determination of Optimal Distances Between Löschian Farmers," *J. A. Zighera*, Centre Francais de Recherche Operationelle

In addition to the presentation and discussion of these formal papers, there will be a series of informal discussions on a variety of topics in Regional Science.

II. Paris

Date: June 14 (3–7 PM) and June 15 (10–12 AM, 3–7 PM)

Place: Institut de Science Economique Appliquee, 35 Boulevard des Capucines

Program: "Statistical Measures of Interregional Relations", *M. Rosenfeld*, Societe D'Economie et de Mathematique Appliquees

"Industrial Complex Analysis and Regional Development", *Walter Isard*, Univ. of Pennsylvania, USA

"Some Consequences of Metropolitan Development in the United States", *Sylvia Fava*, Brooklyn College, USA

"Linking Regional and National Planning", *Philippe Bernard*, Commissariat General de Plan, France

In addition to the presentation and discussion of these formal papers, there will be a series of informal discussions on a series of topics including the following: (1) Statistical Measures of Spatial Relationships, (2) Spatial Models, and (3) Development Programs.

III. Bellagio

Date: July 2 (9–12 AM, 2–5 PM)

Place: Hotel Grande Bretagne

Program: "Techniques of Regional Income Projection", *John H. Cumberland*, EPA

"A Regional Comparison of Economic Structure Based on the Concept of Distance", *Richard Stone*, University of Cambridge, England

"On Urban Development Programs", *J. F. Boss*, Centre Francais de Recherche Operationelle

"A Synthesis of the Interregional Social Accounting System with Other Interregional Systems", *Walter Isard*, University of Pennsylvania, USA

"Interest Rates and Uncertainty in Development Planning", *Otto Eckstein*, EPA

"On Some Transportation Models", *J.R. Boudeville*, Institut de Science Economique Appliquee

IV. Stockholm (in conjunction with the XIXth International Geographical Congress)

Dates: August 6–13

Place: Folkets Hus, Barnhusgatan 14

(*Note*: Specific times and places will be given in the Congress program)

Program: "Regional Science Techniques Applicable to Geographic Studies", *Walter Isard*, University of Pennsylvania, USA

"Some Theoretical Considerations of Macroscopic and Microscopic Aspects of the Geographic Distribution of Income in the United States", William Warntz, American Geographical Society

In addition to the above joint session, an evening meeting of the RSA is planned at which the following papers will be given:

"Some Recent Developments in Interregional Linear Programming", Benjamin H. Stevens, University of Pennsylvania, USA

"Remarks on the Statistical Analysis of Discrete Spatial Distributions", Michael F. Dacey, University of Washington

Various topics in the field of Regional Science will also be discussed.

V. Lund (in conjunction with the Symposium on Problems of Urban Geography)

Date: August 15–19

Place: Department of Geography, Royal University of Lund

(*Note*: Specific times and places will be announced at Lund)

Program: "A Model of the Location of Highway Oriented Retail Business," *Duane F. Marble*, University of Pennsylvania, USA

"Analysis of Central Place and Other Point Patterns by a Nearest Neighbor Method," *Michael Dacey*, University of Washington, USA

"Urban Complex Analysis," *Walter Isard*, University of Pennsylvania, USA

Other Program Participants include:

- Brian J. L. Berry, University of Chicago, USA
- William L. Garrison, Northwestern University, USA
- Edward L. Ullman, Washington University, USA
- Rainer Mackensen, Dortmund, Germany

Regional Science Association, October 20, 1960, Wharton School, University of Pennsylvania, Philadelphia 4, Pennsylvania

Dear Regional Scientist,

In this newsletter I want to report briefly on the various exploratory meetings of the Regional Science Association held in Europe during the last summer. The first meeting was at the Institute of Social Studies, The Hague, Holland, on June 7. The meetings lasted through the morning and the afternoon, and about 35 persons were in attendance. The major focus of discussion was the application of the industrial complex approach in development work.

The second set of meetings took place at the Institute of Applied Economics in Paris on June 14 and 15. These sessions had been very well organized by Professors Perroux and Boudeville. There were sixteen papers presented with some fifteen to twenty discussants, and about fifty persons were in attendance. Emphasis was on regional economic analysis.

The third set of meetings was held at Bellagio, Italy on July 2. There were twenty persons attending and six papers were presented. Attention, in contrast to that of the two previous meetings, centered around more abstract topics. A significant portion of the discussion was concerned with the relations between social accounting on the regional level and the techniques of factor analysis, interregional input-output, comparative costs, and industrial complex analysis.

The fourth set of formal meetings was held in Stockholm on August 11, in conjunction with the meetings of the XIX International Geographical Congress. Approximately seventy-five persons, mostly geographers, were in attendance and four papers were presented. Interregional linear programming was a major topic.

In addition, three informal meetings were held, one at the Institute of Economics of the University of Zagreb on July 11, the second at the Institute of Geography of the Polish Academy of Sciences in Warsaw on July 26, and the third at the Royal University of Lund on August 17 in connection with the Symposium on Problems in Urban Geography. Discussion at the first two of these meetings centered around the use of regional science techniques in planned economies, while the discussion at Lund was related primarily to problems of urban theory and analysis.

The discussions at all of these meetings provided a basis for the exchange of many interesting and stimulating ideas, and considerable interest developed in the activities of the RSA and the new techniques of analysis being evolved by its members. Many persons were eager to become more closely associated with the Association and its work. However, it was recognized that direct participation in the U.S. meetings was impossible for most European scholars. Hence it was suggested by many persons that the Association take several steps to encourage more active participation by Europeans in its work.

One of these suggestions was that there be established in each interested country a section of the RSA which would permit interested persons in the several disciplines within the nation to come together to discuss their research and ideas, and to maintain organized contact with regional scientists elsewhere. Moreover, through these sections it would be possible for members of the RSA in other countries to come into contact with research scholars of similar interests within a given nation. The existence of these sections would also make it easier for members in certain countries to pay their dues in local currencies after appropriate arrangements with the Treasurer of the Association.

In line with this first major suggestion we are submitting for approval at the business meeting to be held during the annual Regional Science Association meetings in St. Louis in December, an amendment to the RSA constitution which would authorize the establishment of RSA sections in foreign countries as well as within the U.S. This latter provision is included because a great deal of interest in the formation of local sections has been expressed by groups in various parts of the U.S.

The second major suggestion concerned the establishment of regular meetings of the RSA in Europe. The purpose of these meetings would be to make it possible for European scholars to learn first hand about the quality research being conducted outside their own country, as well as in disciplines other than their own. I am proceeding to organize the first European meetings, details of which will be reported in a subsequent newsletter.

Walter Isard

P.S.
Please vote on the enclosed ballot and return to: Professor Robert A. Kavesh, Secretary, Regional Science Association, New York University, Graduate School of Business Administration, 90 Trinity Place, New York, NY.

In particular, the interest in regional science in Europe soared at the 1960 Bellagio Conference (organized by the European Productivity Agency (EPA)). After this conference and the other 1960 meetings, the invasion of regional science occurred on a full scale in the thinking of the European scholarly community.

To explain (1) the success of the Bellagio Conference, (2) the wide circulation of the book *Regional Economic Planning: Techniques for Analysis of Less Developed Areas* (1961)[13] covering papers that were delivered, as well as (3) the positive impact resulting from the other 1960 European meetings we must go back to the pursuit of several location studies in the 1950s. The first was the result of the U.S. Department of Commerce research grant to Isard for the study of the development potentials in the Arkansas-White-Red River Basins. This led to a report widely distributed by the Department of Commerce on "Location Factors in the Petrochemical Industry with Special Reference to Future Expansion in the Arkansas-White-Red River Basins" by Walter Isard and Eugene W. Schooler, 1955. Based on the data accumulated in this report, a number of closely related studies were conducted, each contributing to and extending the scope of relevant data. One was "Industrial Complex Analysis and Regional Development with Particular Reference to Puerto Rico" by Walter Isard and Thomas Vietorisz, *Papers and Proceedings of the Regional Science Association*, Vol. I, 1955. Another was "Advantage of Oil Refining" by Robert Lindsay *Papers and Proceedings of the Regional Science Association*, vol. II, 1996, later developed into a doctoral dissertation. A third was the doctoral dissertation, "Location Factors in the Synthetic Fiber Industry with Special Reference to Puerto Rico" by Joseph Airov, 1957, Harvard University. A fourth was Walter Isard and Eugene W. Schooler. 1959. "Industrial Complex Analysis, Agglomeration Economies, and Regional Development," *Journal of Regional Science*, Vol. 1, No. 2, Spring.

Finally, in 1959, all this research, was assembled in the book Walter Isard, Eugene W. Schooler and Thomas Vietorisz, 1959. *Industrial Complex Analysis and Regional Development*, Cambridge, MA: MIT Press. This publication focused on Puerto Rican development and was in large part made possible through significant financial support by the Social Science Research Center, the University of Puerto Rico.

The book was based on: (1) input requirements of diverse oil refineries, and the production of numerous petrochemical products and synthetic fibers, (2) available patent data, (3) best estimates of non-existing data by professors at MIT's Chemical Engineering Department (at that time by far the best in the world), (4) published transport rate data, (5) theoretical transport costs along non-existing routes to and from Puerto Rico estimated by MIT's Department of Naval Architecture and Marine Engineering, (7) proprietary data obtained from chemical construction and engineering firms such as Lummus Chemical, Allied Chemical and Dye Corporation, Foster and Wheeler, Badger and others, where for such data we

[13] Edited by Walter Isard and John H. Cumberland with contributions by Paul N. Rosenstein-Rodan, Alvin Mayne, Richard Stone, Lloyd Rodwin, Otto Eckstein, H. C. Bos and others.

were in a position to offer in exchange our newly developed data, which these firms did not possess.

The research on and underlying the analytical book—which later Theodore Moscoso, Director of the Puerto Rico Development Administration employed to bring to Puerto Rico an oil refinery and the production of various petrochemicals—in effect a major industrial development for this island—was in Europe the payoff for regional science.

In (1) that this book and the subsequent OECD publication embodied location theory's concepts of transport orientation, labor orientation, agglomeration forces and the basic principle that agglomeration strengthens labor orientation, (2) that the economic concepts of comparative cost, scale/spatial juxtaposition/ urbanization economies and externalities were employed (3) that geography's specialization principle was extensively present, (4) that the recent development of regional input-output was used to provide basic underpinning, (5) that the tie with social accounting immediately became obvious, (6) that the research could easily be extended for application using the new tools of linear and nonlinear programming of operations research and management science, (7) that the empirically determined production functions and not the highly abstract ones (e.g. Cobb-Douglas of economics) were employed, as well as *real* data based on engineering experience, (8) that an extensive set of interconnected activities was set forth for use by regional planners and demographers concerned with migration, all led to widespread acceptance of interdisiplinary regional science as a legitimate major area of study.

It should be recalled that in the 1950s and 1960s there was much theoretical and conceptual thinking about economic development. There was the development or growth pole theory of Professor François Perroux[14] and his colleagues. They held that economic development can be best achieved when there is a succession (relatively simultaneous) of balanced growth in a number of interdependent industrial undertakings in a region, which region could then be designated a growth pole. There was also much discussion about the take-off theory of the economic historian Walter Rostow.[15] One interpretation of this theory involving stages of growth holds that when there exists an appropriate institutional structure and a sound cost-revenue basis for operation of one or more leading industrial sectors, investment in them can nurture a dramatic acceleration of the economy.

There followed much criticism of the sketchy empirical materials to support the above and other theories. Also, there was no clear-cut set of past occurrences or a current demonstration of take-off to point to. While Puerto Rico was only an island region and not a nation or large region, the industrial complex study did lead to a

---

[14] Perroux (1956).

[15] Rostow (1956, 1960).

take-off for the island and the resulting oil refinery and associated petrochemical activity did constitute a growth pole of reality. To the many small and undeveloped or underdeveloped regions of Europe and other parts of the world, the Puerto Rico study and the industrial complex and other regional science techniques involved in it generated intense interest in their use in these regions.[16]

To return to the very successful Bellagio conference, it should be noted that it was organized by the *Division for Areas in the Process of Economic Development of the European Productivity Agency* (EPA) and was a response to the urgent requests for study and action addressed by Greece, Italy, Spain, Turkey and Yugoslavia to EPA. These requests were largely spurred on by the regional scientist Edgar S. Dunn, a location theorist trained at Harvard,[17] who as a consultant to EPA "came to perceive the imperative need for a conference and provided the conviction and imagination required to initiate its organization."[18] His vision was further advanced by the diligent work of John H. Cumberland, another young regional scientist trained in location theory at Harvard.[19]

The Bellagio conference proved to be extremely significant for those attendees from underdeveloped regions of Europe—the attendees of the Bellagio conference as well as others who heard about the conference, read the widely circulated book on the conference, and who later had opportunity to read in detail about the new

---

[16] It should be noted that the profitable oil and petrochemicals activities in Puerto Rico came to a halt shortly after the price war of 1973, associated with the newly formed Organization of the Petroleum Exporting Countries (OPEC). As a result, the price of a barrel of oil skyrocketed to as high as $60.00. Because of exceedingly poor management of the Puerto Rican refinery which had arranged for purchase of oil primarily from the spot market and had not established long-term contracts for oil from several sources of supply, costs to produce refinery products became excessive. The refinery could not compete and had to close down. Without the cheap source of refinery products, the petrochemical operations located around the oil refinery became unprofitable and had to close down as well.

Parenthetically this experience pointed up the importance of sound management for the realization of any enterprise that location theory suggests, as well as in the operation of any enterprise in general. This lesson has been learned many times by regional scholars. The history of regional science is replete with instances of poor management: in the closing down of the planning program at the University of Chicago, of geography departments at a number of top universities, of the regional science department at the University of Pennsylvania, etc.

[17] His book on *The Location of Agricultural Production* (Dunn, 1954), based on his doctoral dissertation, constituted a major advance over the von Thünen agricultural location theory.

[18] Isard and Cumberland, 1961, p. 13.

[19] His doctoral dissertation on *Locational Structure of East Coast Steel Industry, with Special Reference to New England* represented by far the best industrial location case study that had been conducted by that time.

techniques of regional science in the book Methods of Regional Analysis published in 1960. They widely opened their doors to regional science. They actively sought the holding of the Regional Science Association's conferences in their countries. And they requested visits of regional science scholars. All further questioning of regional science as a legitimate field of study in effect came to a halt.[20]

The first group actively to seek and arrange for a Regional Science Conference in Europe was the Institute of Social Studies, the Hague, Holland. The lecture given by Isard at this Institute in the month before the Bellagio conference stirred up considerable interest in regional science for planning purposes. Given the further interest generated at Paris and later meetings in 1960 as well as at Bellagio, Professor Jacque Thijsse, a leading regional planner in Europe, was motivated to organize the first of what turned out to be a set of annual European conferences of the Regional Science Association. At the Hague, there were over 120 persons in attendance, largely from European countries. The list of attendees is contained in the Appendix D.

The newsletters of 1961 provide information about the Hague RSA conference; the final program is recorded below. In addition the February 7 newsletter reports on (1) the printing of a Membership Directory of the Regional Science Association, (2) results of the recent election, the officers being:

Honorary Chairman:	W. Isard
President-Elect:	E. M. Hoover
Vice President:	F. S. Chapin Jr.
Councilors:	W. L. Garrison
	J. L. Fisher
	C. Neal
Secretary:	D. F. Marble
Treasurer:	J. Ganschow

(3) a joint session with the Commission on Methods of Economic Regionalization of the International Geographic Union to be held September 7, (4) approval (at the Association's December 1960 business meeting) of the formation of regional sections of the RSA within the United States and of national sections for different countries of the world, (5) the first Western United States meetings of the Association, March 30 to April 1, 1961 in Las Vegas, Nevada, and (6) the scheduling of the annual December meetings (December 27–30) in New York City.

[20] Still later, the annual Papers and Proceedings of the Regional Science Association and the new *Journal of Regional Science* provided much stimulating literature.

Program Regional Science Association
Institute of Social Studies, 27 Molenstraat, Hague, Holland, September 4–7, 1961

(Sessions on September 7 are joint with the Commission on Methods of Economic Regionalization, International Geographical Union)

MONDAY, SEPTEMBER 4, 1961

9:00 A.M. Registration

10:30 A.M. Subject: *Comprehensive Regional and Economic Planning*

Chairman: *E. de Vries*, Institute of Social Studies, Hague

Paper: 1. Planning and Regional Science Techniques
W. Isard and T. Reiner, University of Pennsylvania, USA

Discussants: *P. Bernard*, Paris; *A. Kuklinski*, Institute of Geography, Warsaw

2:30 P.M. Subject: *Studies of Metropolitan Regions*

Chairman: *F. Rosenfeld*, Society of Applied Mathematics, Paris

Papers: 1. The Use of Interchange Models in the Study of Metropolitan Economies
R. Artle, University of California, USA

2. Concept of a Planning Atlas for a Metropolitan Region
A. Kühn, Akademie für Raumforschung und Landesplanung, Germany

Discussants: *J. Piperoglou*, Doxiadis Associates, Athens; *R. Trias Fargas*, University of Barcelona, Spain

TUESDAY, SEPTEMBER 5, 1961

9:30 A.M. Subject: *Industrial Location Studies for a System of Regions*

Chairman: *S. Lombardini*, Instituto Ricerche Economico-Sociali, Torino

Papers: 1. Spatial Distribution of Industry
J. Tinbergen and H. C. Bos, Netherlands Institute of Economics

2. Statistical Explanations of the Relative Shift of Manufacturing Employment among Regions of the United States
V. Fuchs, Ford Foundation, USA

3. The Model of Optimization of Foreign Trade Policy in a Planned Economy
W. Trzeciakowski, Committee on Space-Economy and Regional Planning, Warsaw

Discussants: *P. Holm*, Stockholm; *M. Verhulst*, Paris

2:30 P.M. Subject: *Social Accounting and Income Analysis for Regions*

Chairman: To be announced

Papers: 1. Regional Distribution of National Wealth in Yugoslavia
I. Vinski, Institute of Economics, Zagreb, Yugoslavia

2. Regional Income Distribution in Spain, by Region of Origin and Destination
J. R. Lasuen, University of Barcelona, Spain

Communication: Problems in Constructing an Interregional Input-Output Table for Greece
S. Geronimakis, Ministry of Coordination, Athens

Discussant: *W. Hochwald*, Washington University, USA

WEDNESDAY, SEPTEMBER 6, 1961

9:30 A.M. Subject: *Programming Economic Development in Regions*

Chairman: *J. Thijsse*, Institute of Social Studies, The Hague

Papers: 1. Development Policies for Southern Italy
H. Chenery, Stanford University, USA

2. The Propulsive Industry and the Propulsive Region
F. Perroux, Institute of Applied Economic Science, Paris

Discussants: *J. Boudeville*, University of Lyons, France; *C. Doussis*, Ministry of Coordination, Athens

---

2:30 P.M. Subject: *Spatial Impact of Transport Improvement*

Chairman: *V. Whitney*, University of Pennsylvania, USA

Papers: 1. The Analog of Comparative Method in Economic Geography: Two Models for Predicting Future Use
E. L. Ullman, University of Washington, Seattle, USA

2. Regional Input-Output Model of Shikoku Area and Economic Effect of the Proposed Seto Great Bridge
F. Uyemura, Kagawa University, Japan

Discussion from the floor

5:00 P.M. Meeting of the members of the European sections of the Regional Science Association

---

THURSDAY, SEPTEMBER 7, 1961

(Meetings joint with the Commission on Methods of Economic Regionalization, International Geographical Union)

9:30 A.M. Subject: *Concepts of Region and Regional Structure*

Chairman: *S. Leszczycki*, Institute of Geography, Warsaw, Poland

Papers: 1. Theoretical Problems in the Development of Economic Region
K. Dziewonski, Institute of Geography, Warsaw, Poland

2. Regional Analysis and the Geographic Concept of Region
A. Wrobel, Institute of Geography, Warsaw, Poland

Communication: Regional Input-Output Analysis and Development Programming for Italy
V. Cao-Pinna, Bank of Sicily, Rome

Discussants: *E. Panas*, Ministry of Coordination, Athens; Members of the Commission on Methods of Economic Regionalization of the I.G.U.

2:30 P.M. Subject: *Industrial Complex Analysis*

Chairman: To be discussed

Papers: 1. Large Areal Complexes of Forces in the Soviet Union
J. G. Shaushkin, Moscow University, USSR

Discussants: *M. Verburg*, Economic Technological Institute, Holland; Members of the Commission on Methods of Economic Regionalization of the IGU

Summary Paper: Regional Analysis: Retrospect and Prospect
W. Isard and T. Reiner, University of Pennsylvania, USA

The May 25 newsletter reports upon (1) the formal approval of a Western Section of the Regional Science Association, (2) ongoing discussion on the formation of a West Lakes Section of the Association, and (3) the appointment by the Association's Council of a Special Committee on Summer Training Programs in Regional Science to investigate the possibility of holding a special Summer Institute in 1962.

The August 14 newsletter reports on, among other items: (1) the publication of a Bibliography on Central Place Studies, (2) a Joint Program in City Planning and Regional Science at the University of Pennsylvania, (3) a Symposium on Depressed Areas and (4) a training program on Quantitative Methods in Geography.

The October 2 newsletter reported on (1) the formal approval and establishment of a Western Section, and (2) a meeting to consider the formation of a Midwest Section. The November 22 newsletter covered the following report by Isard.

European Meetings

I wish now to report upon the meetings held at the Institute of Social Studies, Hague, September 4–7, 1961. These meetings were very successful. One hundred and twenty-two persons from twenty-nine different countries were registered. More important, the exchange of ideas in the free and voluminous discussion of the papers was at a high level. We were all exposed to new thoughts and viewpoints, and there was no question in the minds of most participants that the meetings were very worthwhile. I personally found the sessions extremely fruitful in terms of fresh insights which I acquired and in terms of new reactions to and criticisms of the several techniques of regional analysis which I have employed.

These insights and criticisms were largely a result of the emphasis European scholars placed on major region development. The comparison below of the titles of papers given by U.S. scholars and those given by Europeans reveals this.

Hague papers by U.S. scholars:

1. Planning and Regional Science Techniques
2. The Use of Interchange Models in the Study of Metropolitan Economies
3. Statistical Explanations of the Relative Shift of Manufacturing Employment among Regions of the United States
4. The Analog of Comparative Method in Economic Geography: Two Models for Predicting Future Use

Hague papers by non-U.S. scholars (Europeans, except one):

1. Concept of Planning Atlas for a Metropolitan Region Spatial Distribution of Industry

2. The Model of Optimization of Foreign Trade Policy in a Planned Economy
3. Regional Distribution of Natural Wealth in Yugoslavia
4. Regional Income Distribution in Spain, by Region of Origin and Destination
5. Development Policies for Southern Italy[21]
6. The Propulsive Industry and the Propulsive Region
7. Regional Input Model of Shikoku Area and the Economic Effect of the Proposed Seto Great Bridge
8. Theoretical Analysis in the Development of Economic Regions
9. Regional Analysis and the Geographic Concept of Region
10. Large Area Complexes of Forces in the Soviet Union.

Isard also reported that we will be able to collect into a single volume most of the Papers which were presented at the Hague meetings. These will be distributed, we hope, free of charge to all members of the Association.

As a result of the stimulation which the Hague meetings provided, those in attendance urged that the European meetings be held on a regular basis. Accordingly I have now arranged for the next European meetings to be held September 3–6, 1962, at Zürich, Switzerland, in the Aula Room of the Eidgenössische Technische Hochschule. Already I have made considerable progress in arranging for a set of quality papers. I shall inform you of the program details in a later newsletter. Other local information may be obtained from the Conference Secretariat, Regional Science Association, Institut für Ortis-, Regional- und Landesplanung, Eidgenössische Technische Hochschule, Sonneggstrasse 5, Zürich.

Another step in the organizational development of the Regional Science Association was to halt the haphazard issuing of newsletters whenever an occasion warranted. The decision was made starting with 1962, to issue newsletter three times a year; and from 1963 on to issue them in January, May and October.

Also, since newsletters from year 1962 on are increasingly concerned with normal, routine activities of the RSA as it steadily grows in directions already taken, and accordingly increasingly less with new directions, we shall in the ensuing text report only on new developments.

The first newsletter in 1962 (in April) reminds members of the September 3–6 conference to take place at Zurich, Switzerland 1962, the program of which follows.

[21] This paper by the U.S. scholar Hollis Chenery reflects the concern of Europeans since it was the outgrowth of Chenery's long-term research in Italy.

Program Second European Congress Regional Science Association, Aula Room, Eidgenössische Technische Hochschule, Zürich, Switzerland, September 3–6, 1962

MONDAY, SEPTEMBER 3, 1962

9:30 A.M. Registration

10:30 A.M. Subject: *New Techniques in Regional Analysis*

Chairman: To be announced

Paper: 1. Use of Statistical Decision Theory in Regional Planning
Walter Isard and T. Reiner, University of Pennsylvania, U.S.A.

Discussant: *A. Kuklinski*, Institute of Geography, Warsaw, Poland

2:30 P.M. Subject: *Basic Concepts and Goals*

Chairman: *P. Bernard**, Paris, France

Papers: 1. Ekistics and Regional Science
C.A. Doxiadis, Doxiadis Associates, Athens, Greece

2. Regional Planning Objectives: An Attempt to Formulate Their Theoretical Basis
J. Kruczala, Krakow, Poland

Discussants: *J. R. Lasuen*, University of Barcelona, Spain; *G. Manners*, University College of Swansea, Wales; and *J. Friedmann*, Massachusetts Institute of Technology, U.S.A.

TUESDAY, SEPTEMBER 4, 1962

9:30 A.M. Subject: *Problems in Regional Development*

Chairman: *J.R.Boudeville*, University of Lyons, France

Papers: 1. Studies of Underdeveloped Regions in Poland
K. Dziewonski, Institute of Geography, Warsaw, Poland

2. Regional Economic Growth and Industrial Development
M. D. Thomas, University of Washington, Seattle, U.S.A.

Discussants: *M. J. de Meirleir*, Fantus Co., Brussels, Belgium; *S. Lombardini**, Instituto Ricerche Economico Sociali, Turin, Italy; *E. Panas**, Ministry of Coordination, Athens, Greece.

2:30 P.M. Subject: *Statistical Analyses and Studies*

Chairman: *R. Stone**, University of Cambridge, England

Papers:

1. Origins of Per Capita Income Differences Among Regions of Holland
   L. H. Klaassen, Institute of Economics, Rotterdam, Holland
2. Statistical Methods for Spatial Analysis
   R. Bachi, The Hebrew University, Jerusalem, Israel
3. Balance of Payments Studies for the Region of Catalonia
   R. Trias-Fargas, University of Barcelona, Spain

Discussants: *S. Geronimakis*, Ministry of Coordination, Athens, Greece; *A. Wrobel*, Institute of Geography, Warsaw, Poland; *V. Cao-Pinna*, Bank of Sicily, Rome, Italy

---

WEDNESDAY, SEPTEMBER 5, 1962

9:30 A.M. Subject: *Community and Rural Development*

Chairman: *P.E.A. Johnson-Marshall**, University of Edinburgh, Scotland

Papers:

1. A Rural Pattern for the Future
   J. P. Thijsse, Institute of Social Studies, Hague, Holland
2. Simple Models for Community Expansion
   Per Holm, Ekonomisk Planering, Stockholm, Sweden

Discussants: *F. Aberg*, University of Stockholm, Sweden; *T. F. Rasmussen*, University of Oslo, Norway; *Mr. Rallis**, the Technical University of Denmark, Copenhagen

---

2:30 P.M. Subject: *Location Analysis and Theory*

Chairman: *K. H. Olsen**, Akademie für Raumforschung und Landesplanung, Hannover, Germany

Papers:

1. Tendencies in Agricultural Specialization and Regional Concentration of Industry
   M. Chisholm, Bedford College, England
2. Some New Developments of Location Theory
   E. von Böventer, University of Münster, Germany

Discussants: *M. C. Verburg*, Economisch Technologisch Institut Voor Zeeland, Middleburg, Holland; *H. C. Binswanger*, University of Zurich, Switzerland; *I. Friscic**, Institute of Social Sciences, Beograd, Yugoslavia

5:30 P.M. Meeting of the Members of the Regional Science Association

THURSDAY, SEPTEMBER 6, 1962

9:30 A.M. Subject: *Interregional Programming*

Chairman: *C. Salzmann*, Centre Francaise de Recherche Operationnelle, Paris

Papers:
1. Interregional Linear Programming as a Tool for the Analysis of Problems of the Common Market
B. H. Stevens, University of Pennsylvania, U.S.A.
2. The Concept of Accounting Prices in International Trade
W. Trzeciakowski and W. Piaszczynski, Committee on Production Location, Warsaw, Poland
3. A Model of Optimization of Location of Service Stations and Its Practical Applications
J. Mycielski and W. Trzeciakowski, Committee on Production Location, Warsaw, Poland

Discussant: *M. Van der Stichele*, Kortrijk, Belgium

2:30 P.M. Subject: *Urban Interrelations*

Chairman: *J. Servais**, International Center for Regional Planning and Development, Brussels, Belgium

Paper:
1. Structure and Interdependence Among Cities: The Case of Greece and Yugoslavia
B. Ward, Center of Economic Research, Athens, Greece

Discussant: *S. Dabcevic-Kucar**, University of Zagreb, Yugoslavia

Summary Paper: Regional Analysis: Current and Future
W. Isard, University of Pennsylvania, U.S.A.

*Invitation to be confirmed.

In line with the Association's basic interest in urban development, the April 1962 newsletter also called attention to a very useful publication entitled *RESEARCH DIGEST* prepared at the University of Illinois to serve as a medium of communication among individuals and groups engaged in urban and regional research. Additionally, it reported on the results of the association's application to the National Science Foundation for funds to support a summer training program in regional science. A grant in the amount of $49,600 was made jointly to the Association and the University of California to support a unitary institute on the Berkeley campus during the summer of 1962. Funds are available from the grant to support about 35 college and university teachers of social science during an eight-week intensive study period. Full details on the Institute's operations were contained in a brochure, which was mailed to all members of the Association.

7 The Spread of Regional Science into Japan, India, and Latin America

The August 1962 newsletter recorded the spread of deep interest in regional science in Latin America that had developed in previous years. It then announced the *First Latin American Congress of the Regional Science Association*, arranged through the good offices of the new Centro de Estudios del Desarrollo (CENDES), at the Universidad Central de Venezuela in Caracas, Venezuela. The Congress was to take place at CENDES on November 12–14, 1962, and on the following two days Walter Isard was scheduled to conduct a series of open seminars on regional development problems. The theme of the meetings was "Industrial and Regional Development."

Among the participants will be faculty members of CENDES, present and former staff of the United Nation's Economic Commission for Latin America, the Regional Science Research Institute, the Joint Center for Urban Studies (Harvard–M.I.T.), and the regional development groups in Brazil, Venezuela and other nations. As with all Association meetings, the sessions will be open and all members and other interested persons are invited to attend. To enable each participant to speak in the language in which he is most comfortable, simultaneous English-Spanish translation facilities will be available. Anyone who wishes to submit a paper for consideration for presentation at the Congress should write to Professor Walter Isard.

The November 1962 newsletter circulates the Pittsburgh meeting program and specifies publication details for the papers to be presented.

The January 1963 newsletter records that the proposed second Summer Institute will not be held this year due to the lack of financial support. However, the Council indicated that it is in favor of continuing these special training programs if adequate support can be obtained. The Newsletter announces that the policy of distributing the Papers of the Regional Science Association to all dues paying members will be continued with regard to Volume VIII which covers the papers presented at the Hague meetings.

Also, it announces the abandonment of the traditional practice of holding the annual North American meeting at the same time in December as does the American Economic Association (AEA) and other members of the Allied Social Science Associations (ASSA). It was felt that the concomitant sessions of these associations too often attracted scholars who had other specific interests as well as regional ones, and thus interfered with achieving a truly cohesive annual meeting of the RSA. The next annual meeting of our Association, which marks the tenth, is scheduled to be held at the Kellog Center at the University of Chicago, November 15–17.

We present the tentative program of this first independent North American annual conference. Careful study of this program suggests that the most important decision to hold independent annual meetings and to establish the Regional Science Association as an entity itself had no observable negative impact on the quality of papers presented and number of participants.

Tentative Program
Tenth Annual Meetings Regional Science Association
The University of Chicago, Center for Continuing Education, 1307 East 60 Street, Chicago, Illinois November 15–17, 1963

THURSDAY EVENING, NOVEMBER 14

Classical Music Sessions

Organizer: *Dr. Edgar M. Hoover*, Center for Regional Economic Studies, University of Pittsburgh, Pittsburgh, Penna.

FRIDAY MORNING, NOVEMBER 15

8:30 A.M. Registration

10:00 A.M. Subject: *New Concepts and Research Areas in Regional Science*

Chairman: To be announced

Papers:
1. A New Framework for General Interregional Equilibrium Analysis
W. Isard, University of Pennsylvania
2. New Statistical Concepts and Measures for Spatial Analysis
M. Dacey, University of Pennsylvania

Discussant: *W. Warntz*, American Geographical Society

FRIDAY AFTERNOON, NOVEMBER 15

2:30 P.M. Subject: *Classical Location Theory: Some Needed Extensions*

Chairman: *E. M. Hoover*, University of Pittsburgh

Papers:
1. Classical Location Theory Reexamined: Some New Developments
L. N. Moses, Northwestern University
2. On a Social-Psychological Framework for Regional Analysis
M. Webber, University of California

Discussants: *L. Lefeber*, Massachusetts Institute of Technology;
M. Greenhut, University of Florida.

FRIDAY EVENING, NOVEMBER 15

8:00 P.M. Subject: *Transportation Studies*

Chairman: *J. D. Carroll, Jr.*, Tri-State Transportation Committee, New York

Paper: 1. Transportation Models and Analysis: Retrospect and Prospect
D. F. Marble, Northwestern University

Discussants: *J. Meyer*, Harvard University; and *Gary Fromm*, Brookings Institute

---

SATURDAY MORNING, NOVEMBER 16

8:30–9:10 A.M. Early Bird Session I: *Regional Science Ph.D. Theses:*

1. An Input-Output Study for the Calcutta Industrial Region
Manas Chatterji, University of Rhode Island

2. A Model of Urban Growth
S. Czamanski, University of Pennsylvania

9:30 A.M. Subject: *Urban and Land Use Studies*

Chairman: *Robert Campbell*, George Washington University

Papers: 1. Urban Models: A Review and a Forward Look
B. L. J. Berry, University of Chicago

2. On Urban Rent Theory
W. Alonso, Harvard University

Discussants: *L. Wingo*, Resources for the Future, Inc.; *C. Tiebout*, University of Washington, Seattle

---

SATURDAY LUNCH MEETING

Subject: *The Regional Science Ph.D. Curriculum*

SATURDAY AFTERNOON NOVEMBER 16

2:30 P.M. Subject: *Regional Accounting and Input-Output*

Chairman: *H. S. Perloff*, Resources for the Future, Inc.

Papers: 1. Interregional and Regional Accounting in Perspective
C. L. Leven, University of Pittsburgh

2. Some New Types of Regional and Interregional Input-Output Models
W. Leontief, Harvard University

Discussants: *D. Bramhall*, John Hopkins University; and *W. Z. Hirsch*, University of California at Los Angeles

---

5:00 P.M. *Business meeting*

SATURDAY EVENING, NOVEMBER 16

8:00 P.M. *Presidential Address*

Chairman: *W. L. C. Wheaton*, University of Pennsylvania

Address: *W. Garrison*, Northwestern University

SUNDAY MORNING, NOVEMBER 17

9:30 A.M. Subject: *Interregional Programming*

Chairman: *R. L. Ackoff*, Case Institute of Technology

Papers: 1. Interregional Linear Programming: An Evaluation
B. H. Stevens, University of Pennsylvania

2. Statistical Equilibrium and Spatial Organization
R. Vining, University of Virginia

Discussants: *E. O. Heady*, Iowa State University; *H. B Chenery*, Agency for International Development

---

SUNDAY AFTERNOON, NOVEMBER 17

2:30 P.M. Subject: *Resource and Regional Development*

Chairman: *J. L. Fisher*, Resources for the Future, Inc.

Paper: 1. To be announced
J. M. Henderson, University of Minnesota

Discussants: *M. Thomas*, University of Washington, Seattle; *E. S. Dunn, Jr.*, U.S. Department of Commerce; *S. Robock*, Indiana University

---

Approximately 5:00 P.M., *Regional Science JAZZ session*

Organizer: *B. H. Stevens*, University of Pennsylvania

The January 1963 Newsletter also notes that as a result of the successful Zurich conference, a Third European Regional Science Congress will be convened. It will be held at the Royal University of Lund Sweden with local arrangements to be handled by Professor Torsten Hagerstrand, Department of Geography. The program scheduled for August 26–29 follows:

Tentative Program
Third European Congress Regional Science Association
Department of Geography, University of Lund, Lund, Sweden,
August 26–29, 1963

MONDAY, AUGUST 26, 1963

9:30 A.M. Registration

10:30 A.M. Subject: *New Concepts for Spatial Organization and Regional Planning*

Chairman: *T. Hagerstrand*, University of Lund, Sweden

Paper: 1. Spatial Decentralization of Decision-Making and Planning in a System of Regions
W. Isard, University of Pennsylvania, U.S.A.

Discussants: *F. Rosenfeld*, Societe d'Economie et Mathematique Appliquees, Paris; *A. Wrobel*, Institute of Geography, Warsaw

2:30 P.M. Subject: *Studies on Community Growth and Development*

Chairman: *O. E. Koenigsberger*, School of Architecture, London

Papers: 1. Simple Models for Community Expansion
P. Holm, Ekonomisk Planering, Stockholm

2. Factors Influencing Suburban Land Development
S. J. Maisel, University of California (Berkeley), U.S.A.

Discussants: *J. Maurer*, Zurich; *R. Mackensen*, Social Research Center, Dortmund

TUESDAY, AUGUST 27, 1963

9:30 A.M. Subject: *Location Theory and Studies*

Chairman: *S. Dahl*, Gothenburg School of Economics and Business Administration, Sweden

Papers: 1. Some Aspects of Location Analysis
M. C. Verburg, Economisch Technologisch Instituut voor Zeeland, Netherlands

2. Industry Location Factors
M. J. de Meirleir, Fantus Company, Brussels

Discussants: *R. Funck*, Institut für Verkehrswissenschaft, Münster; *E. Wiren*, Arkitekt och Ingeniorskontoret, Stockholm

---

2:30 P.M. Subject: *Central and Peripheral Place Theory*

Chairman: *J. Thijsse*, Institute of Social Studies, The Hague

Papers: 1. The Spacing of Central Places in Sweden
G. Ollson and A. Perrson, Uppsala University, Sweden

2. Some Considerations of Tourism Location in Europe
W. Christaller, Judenheim, Germany

Discussants: *K. Imberg*, Eidgenössische Technische Hochschule, Zürich

---

WEDNESDAY, AUGUST 28, 1963

9:30 A.M. Subject: *Spatial and Regional Models*

Chairman: To be announced

Papers: 1. A Linear Programming Model for the Coastal Region of Northeast Norway
J. Serck-Hanssen, University of Oslo, Norway

2. Some Thoughts on Spatial Models for Development Purposes
J. Sebestyen, Institute for Farm Economics, Budapest

Discussants: *A. Kuklinski*, Institute of Geography, Warsaw; *E. Olsen*, Institute of Economics, Copenhagen

---

2:30 P.M. Subject: *Statistical Studies for Regional Analysis*

Chairman: *J. Fisher*, Resources for the Future, Inc., Washington, D.C.

Papers: 1. Regional Growth of Fixed Assets in Yugoslavia, 1946–1960
I. Vinski, Ekonomski Institut, Zagreb

2. To be announced

Discussants: *G. Fossi*, Organization for Economic Cooperation and Development, Paris; *D. Fillinger*, University of Berne, Switzerland

5:00 P.M. Meeting of the members of the Regional Science Association

THURSDAY, AUGUST 29, 1963

9:30 A.M. Subject: *Transportation Theory and Analysis*

Chairman: To be announced

Papers: 1. The Location of Transportation Routes
W. L. Garrison, Northwestern University, U.S.A.

2. The Impact of the Development of Transportation on the Optimal Size of Plants and on Regional Location
C. Kadas, Faculty of Transport Engineering, Budapest

Discussants: *E. von Böventer*, University of Münster, Germany; *B. Horvat*, Federal Institute of Economic Planning, Belgrade

2:30 P.M. Subject: *Regional Development*

Chairman: *E. L. Ullman*, University of Washington, U.S.A.

Paper: 1. Regional Resource Relocation Problems in a Developing Economy
G. Coutsoumaris, Center of Economic Research, Athens

Discussants: *K. A. Boesler*, Geographisches Institut, Berlin

Summary Paper: Regional Analysis: Needed Developments
W. Isard, University of Pennsylvania, U.S.A.

The January 1963 newsletter also announces that the annual meetings of the Western Section of RSA are to be held at the University of Oregon on June 14–15, sponsored by the Department of Economics.

Importantly, it notes another major expansion of the domain of the Regional Science Association. At its last meeting the Council approved the formation of the Japan Section of the RSA, the first to be established outside of North America. The new officers of the Japan Section are:

President:	Dr. Genpachiro Konno, Tokyo University
Vice-President:	Dr. Eiji Kometani, Kyoto University
Directors:	Dr. Masaji Suzuki, Nippon University
	Dr. Yoshikatsu Ogasawara, Ministry of Construction
	Dr. Susumu Kobe, Waseda University
	F. Uyemura, Kagawa University

A second new section established by the Council was the Midwestern Section designed to cover the states of Illinois, Indiana, Iowa, Michigan, Minnesota, Missouri, Ohio, and Wisconsin. The new officers of the Midwestern Section are:

President:	Dr. James Henderson, University of Minnesota
Vice-President:	Dr. Leon Moses, Northwestern University
Secretary:	Dr. John D. Nystuen, University of Michigan

The proposed Southeast section of the Regional Science Association held an organization meeting at Atlanta, Georgia on November 9, 1962. At that meeting it was decided to petition the Council of the Regional Science Association to establish the Southeast section of the Regional Science Association. (The proposed constitution was subsequently approved by the Council). The meeting at Atlanta involved a particularly fruitful interchange among the participants on the relative merits of (1) a statistical equilibrium approach to the analysis of spatial organization, and (2) an optimizing approach. Because of the interesting exchange of views it was decided to distribute, at a cost of $1.00, the set of papers which were presented.

The newsletter also notes the success of the Caracas conference around the theme of planning and decision making and the investigation of the possibility of holding a second Latin American conference.

The May 1963 newsletter notes that (1) the annual meeting of the French Regional Science Association will be held in Bordeaux on June 7-8, the main topic of which will be "Economic Accounting and Natural Resources," (2) a Greek Section is in process of formation, and (3) efforts at securing financial support for a second Summer Institute in Regional Science are continuing.

It reported that the *First Latin American Regional Science Congress* saw nearly 100 people take part in the three days of meetings in Caracas last November. Of these, about 60 were Venezuelan participants, and 30 came from abroad, the largest delegation being from Puerto Rico. Most participants (aside from the CENDES student body and faculty) were actively engaged in government planning, economic development, and similar tasks. Their major interest was in determining which techniques are operational, and how they can be used in public decision making. In the informal discussion in and around the Congress, much heat (and some light) was generated by questions such as "What is a region's due share in national investment?" and "How can a deprived region make its demands felt before a national body?"

The May 1963 newsletter also mentioned that as the Tenth annual meeting in Chicago was a meeting independent of the American Economic Association and others of the Allied Social Science Associations, it was possible to plan to enliven the meetings with special musical festivities. As noted in the tentative Chicago program, on Thursday evening (November 14) one or more classical music sessions will be arranged, and the meetings will close with a jazz session immediately following the last discussion on Sunday (November 17). Members were strongly encouraged to participate in these activities, and all members who wish to do so are urged to write directly to the session organizers as soon as possible. These organizers are:

Classical sessions:	Edgar M. Hoover, Center for Regional Economic Studies, University of Pittsburgh, Pittsburgh, Penna.
Jazz session:	Dr. Benjamin H. Stevens, Department of Regional Science, Wharton School, University of Pennsylvania, Philadelphia 4, Penna.

While the classical sessions did not continue for long at the annual meetings, the jazz sessions continued almost regularly at annual North American meetings. The latter outcome was largely due to the efforts and inspiration of Benjamin H. Stevens who with enthusiasm arranged jazz sessions until his most unfortunate death in early December, 1997.

On May 30, 1963 a special newsletter notified members of the recently arranged first Far East Conference in Tokyo on September 12–14 (with an additional day of seminars on the 16th). The Japan Section of the Regional Science Association will be the host of this conference. As with every Regional Science Association conference, members of the Association, their associates and other interested individuals are most welcome to attend. Those who wish to present papers or serve as discussants were asked immediately to inform W. Isard. The program of the First FAR East Conference at Tokyo was:

First Far East Conference
R.S.A. Tokyo, Japan, September 11-14, 1963

WEDNESDAY, SEPTEMBER 11, 1963

9:30 A.M.	Registration
10:30 A.M.	Words of Welcome: *G. Konno*
10:40 A.M.	Subject: *New Concepts and Models in Regional Science*
Chairman:	*S. Kobe*, Waseda University, Japan
Paper: 1.	Spatial Organization and Administration for National and Regional Planning *W. Isard*, University of Pennsylvania, U.S.A.
Discussant:	*T. Sasada*, Doshisha University, Japan

2:00 P.M.	Subject: *Regional Growth Problems*
Chairman:	*S. Ohkita*, Economic Planning Agency, Japan
Papers: 1.	The Normal Requirements Technique for Estimating the Future Work-Force of Towns—with special Reference to Australia *G. J. R. Linge*, The Australian National University, Australia
2.	Urban Aspects of Indian Development *J. E. Stepanek*, SIET Institute, Hyderabad, India
Discussants:	*C. Yoshida*, The Bureau of Capital City Development, Tokyo, Japan; *T.A. Reiner*, University of Pennsylvania, U.S.A.

THURSDAY, SEPTEMBER 12, 1963

9:30 A.M.	Subject: *Location Theory and Regional Analysis*
Chairman:	*F. Uyemura*, Kagawa University, Japan
Papers: 1.	Morphology and Economic Theory of the Industrial Agglomeration *D. Esawa*, Senshu University, Japan
2.	A Model of Regional Planning *S. Ichimura*, University of Osaka, Japan
Discussants:	*H. Aoyama*, University of Kyoto, Japan; *Z. Ito*, Tokyo Women's Christian College, Japan

1:30 P.M.	Subject: *Urban Studies*
Chairman:	*K. Tange*, University of Tokyo, Japan
Papers: 1.	Housing Stock as Resources *Wallace F. Smith*, University of California (Berkeley), U.S.A.
2.	Some Characteristics of Community and Regional Structure in Japan *E. Isomura*, Tokyo Metropolitan University, Japan
3.	Some Geographic Aspects of Urbanization and Metropolitanization in Japan *S. Kluchi*, University of Tokyo, Japan
Discussants:	*R. Nagata*, Japan Housing Corporation, Japan; *K. Baba*, Economic Planning Agency, Japan; *H. Sazanami*, Ministry of Construction, Japan

---

FRIDAY, SEPTEMBER 13, 1963

9:30 A.M. Subject: *Physical Planning and Transportation*

Chairman: *S. Ichimura*, University of Osaka, Japan

Papers: 1. Model Plan of a New Industrial Area Through Interactivity Matrix
E. Kometani, University of Kyoto, Japan

2. A Microscopic Theory of Traffic Assignment
N. Sakashita, University of Tohoku, Japan

Discussants: *M. Ohta*, Waseda University, Japan; *F. Okazaki*, Ritsumeikan University, Japan

---

1:30 P.M. Subject: *Regional Development Policies in the Pacific Area*

Chairman: *Y. Ogasawara*, Ministry of Construction, Japan

Papers: 1. Regional Development Policy in Japan
S. Ohkita, Economic Planning Agency, Japan

2. Development Strategy for Northern and Western Alaska
Hubert J. Gellert, University of Alaska, U.S.A.

3. Regional Development in the Philippines
Angel Q. Yoingco, Joint Legislative-Executive Tax Commission, Philippines

4. Regional Science and Economic Development; The Southeast Alaska Region
C. Rogers, University of Alaska, U.S.A.

Discussants: *S. Sakai*, University of Nagoya, Japan; *G. Konno*, University of Tokyo, Japan; *Y. Ohishi*, University of Tokyo, Japan; *S. Hara*, Economic Planning Agency, Japan

---

5:30 P.M. Cocktail Party at Toshi Center Dining Room

SATURDAY, SEPTEMBER 14, 1963

9:00 A.M. Round Table Seminar
W. Isard, University of Pennsylvania, U.S.A.; *T.A. Reiner*, University of Pennsylvania, U.S.A.

The September 1993 newsletter announces the publication of an English edition of the *Papers of the First Latin American Conference on Regional Science.*

Preface by the English Editor

General Papers:

1. Walter Isard and Thomas Reiner:
   Analytic Techniques for National and Regional Planning
2. Jorge Ahumada, Luis Lander, Eduardo Neira Alva:
   Latin America: A Challenge to Regional Science

Venezuela:

3. John Friedmann:
   Economic Growth and the Urban Structure in Venezuela
4. Jorge A. Casanova:
   Some Considerations Concerning the Process of Development in a Multiregional Spatial Setting
5. Alexander Ganz:
   Regional Planning as a Key to the Present Stage of Economic Development of Latin America: The Case of the Guayana Region, A Frontier Region

Latin America:

6. Alberto Fracchia, Norberto Gonzalez, Hector Grupe and Felipe Tami:
   Application of Quantitative Economic Analysis to the Programming of Regional Development: An Experience of the Argentine Republic
7. Mary Megee:
   Factor Analysis Applied to Mexico

Regional Technics:

8. Thomas Reiner:
   Regional Investment Allocation Criteria
9. Leon N. Moses:
   Allocating Investment among Regions

A Spanish edition is also being printed.

The Newsletter of November 1963 reported that the *Fourth European Regional Science Congress* will be held at the State University of Ghent, Ghent, Belgium, July 14–17, 1964. Under the sponsorship of the Dutch speaking section of the Regional Science Association (now in process of formation), the Seminar for Applied Economics (SERUG) directed by Professor A.J. Vierick will serve as host. Information can be obtained by writing to the Secretariat of the Fourth European Regional Science Congress, Seminar for Applied Economics, State University of Gent, Universiteitstrat 16, Gent, Belgium. Details of the program will be in forthcoming newsletters.

An innovation at this Congress will be an organized program for wives who accompany their husbands. Also, informal classical music sessions will be organized for those of us who wish to participate actively in music-making. Interested persons should write well in advance to Professor Jac. P. Thijsse, Institute of Social Studies, 27 Molenstraat, the Hague, Holland.

The November 1963 also announces that a *Second Summer Institute* in regional science will be held from June 14 through July 27 of this year on the Berkeley campus of the University of California. This post-graduate training program is carried out jointly by the University of California and the Association under the terms of a grant from the National Science Foundation. This year, stipends, dependency allowances, and modest travel grants will be available to 40 college and university teachers who wish to take part in the program. In addition, the University of California has arranged to accommodate five or six additional persons (non-college teachers) on a tuition basis. All inquiries about participation should be directed to Dr. Duane F. Marble, Department of Geography, Northwestern University, Evanston, Illinois 60201. Once again, participation by foreign teachers is welcomed.

The February 1964 newsletter contains the following announcements:

Eleventh Annual Meeting of the Regional Science Association, Ann Arbor, November 13–16, 1964

Because of the high level of interaction, exchange of ideas, and sociality which was attained at the Tenth Annual Meeting of the Regional Science Association at the University of Chicago, November 1963—which was the first time the Regional Science Association met independently of the Allied Social Science Associations—the members of the Council voted to hold the Eleventh Annual Meeting as another independent conference. The Eleventh Annual Meetings are now scheduled to be held at the Michigan Union, University of Michigan, Ann Arbor, Michigan, Friday, November 13 to Monday, November 16. Professor John Nystuen of the Department of Geography of the University of Michigan has agreed to serve as local arrangements chairman.

Regional Science Student Organization, Harvard University and M.I.T.

Readers of this newsletter may be interested to know that graduate students in city and regional planning, economics, sociology, and other social science fields at Harvard University and M.I.T. have set up an informal organization in order to learn more about regional science as a field of study. This organization, with the advice of the Regional Science Research Institute, is arranging a series of introductory lectures by outstanding regional scientists and other social science scholars.

Faculty of Urban Studies and Regional Science, Washington University, St. Louis

In keeping with its strong interest in urban studies and regional science, Washington University has appointed a faculty on urban studies and regional science with the following membership: *Joseph R. Passonneau, Chairman*, Harold Barnett, David B. Carpenter, Leon Cooper, Alvin W. Gouldner, Laurence Iannaccone, Daniel R. Mandelker, Carl McCandless, Roger Montgomery, David Pittman, Lee Robins, Robert Salisbury, Wayne Vasey, and Sam Warner. Relevant graduate courses are currently given in the departments of architecture, economics, education, engineering, history, law, political science, sociology, and social work. Two educational programs are offered. One is a post-graduate, non-degree program, open to students who have taken their Ph.D. or have passed general examinations for the Ph.D., or have taken the M.A. in urban design, the M.A. in social work, the L.L.B., or the D.B.A.

The other program in urban studies and regional science is part of the individual's degree course work toward a Ph.D. in one of the departments, or toward the M.A. in design, or the L.L.B., or the D.B.A., or the M.A. in Engineering.

Ad Hoc Committee on Classifications for Regional Input-Output Studies

It may be of interest to the membership to know that the Regional Science Research Institute has set up an ad hoc committee on classifications for regional input-output studies. This committee consists of Wassily Leontief, Harvard University, Chairman; David Bramhall, Johns Hopkins University; John Carden, Mississippi Industrial and Technological Research Commission; John Cumberland, University of Maryland; Morris Goldman, Office of Business Economics, U.S. Department of Commerce; William Miernyk, University of Colorado; Charles Tiebout, University of Washington; Harold Barnett, Washington University; William Garrison, Northwestern University; Werner Hirsch, University of California at Los Angeles; Leon Moses, Northwestern University; and Walter Isard, Regional Science Research Institute, Secretary.

The objective of this committee is to develop a system of classification of industries, primarily, and regions, secondarily, in order to facilitate comparison of the results of different regional input-output studies. This should make it possible for an analyst to use the empirical findings and results of other analysts in constructing input-output tables for his own region and in developing interregional models. At a later time, members of the R.S.A. will receive a set of recommendations of this committee and will be informed of its work.

French Language Regional Science Association Meetings.

We have received notice that the next regional congress of the Association de Science Regionale de Langue Française will be held in Liege, Belgium, May 21 and 22, 1964, under the direction of Professor Louis Davin.

Selected Minutes of the Business Meeting at the Tenth Annual Meeting of the Regional Science Association

The meeting was called to order at 5:00 P.M. by William Garrison, the President.

Items of Business

1. Adoption of minutes of last year's Business Meeting was approved
2. Treasurer's Report

   Benjamin Stevens reported informally for the absent treasurer. He noted that at the present the Association has 1900 members, 650 outside the United States. With higher dues, for the moment at least, the Association has a sizeable balance in the Treasury.

   President Garrison asked about the impact of the higher dues on membership, which brought forth the information that few members seem to have dropped out and perhaps even higher dues can be set in the future.
3. New Sections and Foreign Meetings

   Walter Isard stated that:

   a. The Greek Section has an application before the Council for review. There are a number of questions to be raised about the submitted constitution.
   b. The application of the Scandinavian Section is in good order.
   c. The Dutch-speaking members are in the process of writing up a constitution.
   d. The Japanese constitution submitted last year was approved.

Other groups of scholars who have definite interest in forming sections include the British, Italian, Spanish, and the Philippine.

Walter Isard also described a visit with Thomas Reiner to Russia; and pointed out that two papers by members of the Academy of Science, Moscow were to be given at the Ghent meetings.

Professor Genpachiro Konno, President of the Japanese section, was present at the Chicago meetings and was introduced. He said that he hoped that there would be a second Far East conference in 1965; and that the Far Eastern members would be very obliged to American members who come.

Finally, Isard indicated that a New England group met and submitted a constitution for approval.

The April 1964 Newsletter mentioned that at the forthcoming Fourth European Regional Science Congress at Ghent, Belgium, July 14-17, there will be, at the request of scholars at the 1963 Lund Congress, some emphasis on problems of the Common Market. Also, information was provided on the International Geographic Union meetings in London, England, July 20-28, and on the meetings and papers of the International Peace Research Society to be held at Ghent on July 18-19. The Ghent program follows:

Tentative Program
Fourth European Congress Regional Science Association
Seminar for Applied Economics, State University of Ghent,
Universiteitstraat 16, Ghent, Belgium
July 14–17, 1964

TUESDAY, JULY 14, 1964

9:30 A.M. Registration

10:30 A.M. Subject: *Regional Theory*

Chairman: *A. J. Vlerick*, Seminar for Applied Economics, Ghent

Paper: 1. Some New Aspects of General Interregional Theory
W. Isard, University of Pennsylvania, U.S.A.

Discussant: *W. J. Lissowski*, Central School of Planning and Statistics, Warsaw, Poland

2:30 P.M.	Subject: *Analysis of the Common Market Economy I*
Chairman:	*M. Verburg*, Economisch Technologisch Instituut Voor Zeeland, Netherlands
Papers: 1.	Regional Income Differences Within a Common Market *E. Olsen*, University of Copenhagen, Denmark
2.	The Impact of the Common Market on Industrial Location and Transport Flows *R. Funck*, Technische Hochschule Karlsruhe, Germany
Discussants:	*J.B.D. Derksen*, Netherlands Central Bureau of Statistics, The Hague; *H. S Perloff*, Resources for the Future, Inc., Washington, D.C.

WEDNESDAY, JULY 15, 1964

9:30 A.M.	Subject: *Mathematical Models in Programming Development*
Chairman:	*L. H. Klaassen*, Netherlands Economics Institute, Rotterdam
Papers: 1.	A Model of Interdistrict Relationships in a Single System Optimum Plan of the Economy *V. S. Dadayan*, Academy of Science, Moscow, U.S.S.R.
2.	Application of Decompositional Methods in the Economic Planning of Developing Countries *T. Kronsjö*, Institute of National Planning, Cairo, U.A.R.
Discussants:	*E. von Böventer*, Heidelberg University, Germany (one other to be announced)

2:30 P.M.	Subject: *Regional Research and Development Planning: the Case of East-Flanders*
Chairman:	*J. Thijsse*, Institute of Social Studies, The Hague, Netherlands
Papers: 1.	General Presentation *A. J. Vlerick*, Seminar for Applied Economics, Ghent
2.	The Economist's Contribution *J. Maton*, Seminar for Applied Economics, Ghent
3.	The Objectives and Approach of the Sociologist *A. Buyst*, Seminar for Applied Economics, Ghent
4.	The Interdisciplinary Approach as a Basic Need *M. Anselin*, Seminar for Applied Economics, Ghent

Discussants: *H. Esping*, University of Lund, Sweden; *R. P. Wolff*, Coral Gables, Florida, U.S.A.

THURSDAY, JULY 16, 1964

9:30 A.M. Subject: *Regional Planning Policy*

Chairman: *S. Geronimakis*, Ministry of Coordination, Athens, Greece

Papers:
1. Coordination of Regional and National Development Policy
   A. Kiiskinen, Nekala, Finland
2. An Allocation Model for Regional Investment
   T. Reiner, University of Pennsylvania, U.S.A.

Discussants: *E. G. Panas*, Ministry of Coordination, Athens, Greece; *G. Fossi*, O.E.C.D., Paris, France

---

2:30 P.M. Subject: *Transportation Analysis and Projection*

Chairman: *R. Mackensen*, Sozialforschungsstelle an der Universität Münster, Dortmund, Germany

Papers:
1. Concepts and Approaches for Traffic Forecasting
   E. Sylven, Goteborgs Generalplanekontor, Sweden
2. Some Reflections on an Urban Transportation Model
   J. F. Kain, U.S. Air Force Academy, Colorado

Discussants: *B. Garner*, University of Leeds, England; *M. Megee*, Washington University, St. Louis, U.S.A.

5:00 P.M. Meeting of the members of the Regional Science Association

---

FRIDAY, JULY 17, 1964

9:30 A.M. Subject: *Location and Regional Analysis*

Chairman: (to be announced)

Papers:
1. On the District Interbranch Balance of Output Production and Distribution in the U.S.S.R.
   V. V. Kossov, Academy of Sciences, Moscow, U.S.S.R.
2. Networks and the Location of Economic Activities
   R. Lächene, Societe D'Economie et de Mathematique Appliquees, Paris, France

Discussants:	*P. Müller*, Institut für Agrarpolitik und Sozialökonomik des Landbaus, Stuttgart, Germany; A. Wrobel, Institute of Geography, Warsaw, Poland
2:30 P.M.	Subject: *Analysis of the Common Market Economy, II*
Chairman:	*A. Kuklinski*, Institute of Geography, Warsaw, Poland
Paper:	(to be announced)
Discussant:	(to be announced)
Summary Paper:	*W. Isard and T. Reiner*, University of Pennsylvania, U.S.A.

The April 1964 newsletter also announced that the Fifth European Regional Science Congress will be held in Warsaw, Poland, September 1–4. The venue and dates were later changed to Cracow, August 30–September 3.

The September 1964 newsletter reminded members of the forthcoming Fifth European Congress to be held in Poland at which it is anticipated that there will be participation of scholars from all continents, and that the forthcoming Eleventh U.S. Annual Meeting, November 13–16, University of Michigan would have both classical music and jazz sessions.

The newsletter also announced:

British Isles Regional Scientists

On July 11, 1964, a meeting was held to consider initial steps in the formation of a British Section, Regional Science Association. Some sixty scholars from various social science fields participated in this meeting, which was held at the London School of Economics, under the chairmanship of Professor Michael Wise. Those interested in this endeavor should contact Dr. G. P. Hirsch, Agricultural Economics Research Institute, Oxford University, Oxford, England, who is the Secretary of the Steering Committee.

German-Speaking Committee on Regional Science

German and Swiss regional scientists, some 30 strong, met for two weeks of informal meetings during April 1964. In addition to several interesting papers, the participants also joined in various activities such as skiing. Dr. Edwin von Böventer, Alfred-Weber-Institut, University of Heidelberg, Heidelberg, Germany, one of the organizers of this endeavor, assures us that this procedure will be repeated during the coming year.

Scandinavian Meeting

Regional scientists from Scandinavia met during the spring at the University of Lund, Sweden. A number of papers were circulated analyzing the effects of various proposals for a bridge connecting Sweden and Denmark.

New England Section, Regional Science Association

Professor John Friedmann, outgoing President, reports that the 1964 Spring Meeting of the Regional Science Association—New England Section, was held on April 11, 1964 at the A.D. Little Company in Cambridge, Mass. Approximately 80 people were in attendance. George Borts of Brown University was elected President for 1964-65; and Peter Stern of the A.D. Little Company, Vice President. Edward Gooding of the Federal Reserve Bank, Boston, Mass. continues as Secretary.

The following papers were presented and discussed:

1. Louis Lefeber (MIT), "Regional Economic Growth and National Development." Discussant: Raymond Vernon (Harvard).
2. George Lewis (Boston U.), "Retail Store Location: Theory and Practice." Discussant: Bernard Frieden (MIT).
3. James Nelson (Amherst), "Comparative Advantages and Decreasing Costs: The Special Case of New England Transport." Discussant: John Meyer (Harvard).

Results of an inquiry on current regional research in New England were distributed at the meeting.

The March 1965 newsletter recorded:

Second Far East Regional Science Conference

We have just received word that the second Far East Regional Science Conference is now definitely scheduled for Tokyo, September 15-17, 1965. The host for this conference will be the Japanese Section, Regional Science Association, of which Professor G. Konno is President. The meetings will be held at Toshi Center, 6 Hirakawa-cho, Chiyoda-ku, Tokyo. The program will be organized in much the same fashion as was the first Far East Conference. (See Newsletter, November 1963, for program details.) Although the conference is designed to stimulate and further the development of research and training programs in regional science among scholars of the Far East, papers by regional scientists from other parts of the world are welcome.

The scheduling of Second Far East conference in Japan testified to the fact that the First Far East conference was exceedingly successful, and together with the First, it sparked the rapid and extensive development of regional science in Japan and its hinterland in Asia.

New England Section, Regional Science Association

The spring meeting of the New England Section, Regional Science Association, is scheduled for April 3, 1965, at Jeppson Hall, Clark University, Worcester, Massachusetts. Papers will be given by Maynard Hufschmidt (Harvard University), David Major (Harvard University) and Robert Kates (Clark University).

Regional Science Graduate Student Organization, M.I.T. and Harvard

The Regional Science graduate students organization of M.I.T. and Harvard held the fourth in its series of 1964-65 lectures on Wednesday, March 17, 1965, at Hunt Hall A, Harvard University. Dr. Charles Tiebout, Director of the Center of Regional Studies and Professor of Economics at the University of Washington, spoke on *The Urban Consumption Function and the Economic Base.*

The March 1965 newsletter also announced that the Twelfth U.S. Annual Meeting of the Regional Science Association will be held at the Annenberg School of Communications, University of Pennsylvania, Philadelphia, Pa., Friday, November 12 through Sunday, November 14. It recorded the following items of the November 1964 Business Meeting:

Report of Data Committee

David Bramhall reported that (1) an ad hoc committee of the Regional Science Research Institute is continuing to consider the matter of standardization of industrial classification in regional input-output studies; and (2) the Data Committee had discussed the matter of standardization of classification of geographical areas, and decided to circulate a questionnaire to the members of the Association.

Report on Foreign Sections

Walter Isard reported that the constitution for the Greek Section, Regional Science Association, was being revised; that the Council had approved the Scandinavian constitution; and that a group was at work in Great Britain organizing a British Section.

The *June 1965* newsletter presented the tentative program of the *Fifth European Regional Science Congress*, which is to take place in Cracow, Monday, August 30, through Friday morning, September 3. The Committee for Space Economy and Regional Planning of the Polish Academy of Sciences is the Polish host of the Congress. The Polish Town Planning Society has kindly offered its assistance in the preparation of local arrangements in Cracow.

As indicated in the previous Newsletter, the Polish hosts have established a special Honorary Committee to sponsor the Congress. The Chairman of the Honorary Committee is Professor S. Leszczycki, the President of the Committee for Space Economy and Regional Planning. The members of the Honorary Committee are

Professor M. Klimaszewski, Rector of the Jagiellonian University in Cracow, Mr. Z. Skolicki, the Mayor of the City of Cracow, and Professors M. Kaczorowski, B. Pniewski, K. Secomski and A. Kuklinski, members of the Executive Board of the Committee for Space Economy and Regional Planning. Dr. Jerzy Kruczala has been appointed the Executive Secretary of the Honorary Committee.

In the previous Newsletter were enclosed Form A (for room reservation) and Form B (for administrative authorities in Cracow) to be filled out by all those who plan to attend. The forms are to be sent air mail to Dr. J. Kruczala, Cracow, Pijarska 9, as soon as possible. There will be no special visa arrangements for the participants of the Congress. Participants are kindly requested to follow the standard procedure and apply for a visa in the proper Polish Embassy or Consular office. In case you have any special problem concerning your visit to Poland, please write to Dr. A. Kuklinski, the Executive Secretary of the Committee for Space Economy and Regional Planning, Warsaw 64, Krakowskie Przedmiescie 30.

Note that the Friday morning (September 3) session is jointly sponsored with the Peace Research Society (International) which is holding its Second European Congress in Cracow, September 3-5, 1965. All interested persons are welcome to attend this Congress. A tentative program can be obtained by writing to the Peace Research Society (International), Department of Regional Science, Wharton School, University of Pennsylvania, Philadelphia 4, Pennsylvania, U.S.A.

Tentative Program
Fifth European Congress Regional Science Association
Cracow, Poland, August 30–September 3, 1965

Host: Committee for Space Economy and Regional Planning, Polish Academy of Sciences

Local Arrangements Secretary: Dr. J. Kruczala, Pijarska 9, Cracow, Poland

MONDAY, AUGUST 30, 1965

9:30 A.M. Registration

10:30 A.M. Subject: *Regional Science as a Field of Study: A Reexamination*

Chairman: *S. Leszczycki*, Committee for Space Economy and Regional Planning, Warsaw

Paper: 1. Regional Science: Retrospect and Prospect
W. Isard and T. A. Reiner, University of Pennsylvania, U.S.A

Discussant: *M. Anselin*, Seminar of Applied Economics of Ghent, Belgium

4:00 P.M. Subject: *Reports on Regional Science Activities in Diverse Regions of the World*

Papers:

1. Regional Science in Poland
   S. Leszczycki and A. Kuklinski, Committee for Space Economy and Regional Planning, Warsaw
2. The Japanese Regional Science Section
   G. Konno, University of Tokyo, Japan
3. The Scandinavian Regional Science Section
   T. Hägerstrand, University of Lund, Sweden
4. Regional Science Activities in Africa
   B. Kwiatkowski, Town and Country Planning Division, Accra, Ghana
5. The French Speaking Regional Science Organization
   J. Boudeville, Faculty of Law and Economics of Paris, France

---

TUESDAY, AUGUST 31, 1965

9:30 A.M. Subject: *Integration Processes and Regional Problems*

Chairman: *A.J. Vlerick*, Seminar for Applied Economics of Ghent, Belgium

Papers:

1. Problems of Integrated Planning for the Country Members of COMECON
   A Bodnar, Central Committee of the Polish United Workers party, Warsaw
2. Industrial Relationships and Location Analysis for a System of Regions
   L. Klaassen, Netherlands Economic Institute, Rotterdam.

Discussant: *R. Lachene*, Society of Applied Economics and Mathematics, Paris

---

4:00 P.M. Subject: *Optimal Growth Patterns: National and Regional*

Chairman: *C. Kadas*, Faculty of Transport Engineering, Budapest

Papers:

1. Regional Models and National Economic Planning
   K. Porwit, National Planning Commission, Warsaw
2. (to be indicated)

Discussant: *V. Rajkovic*, Republic Institute for Economic Planning, Zagreb, Yugoslavia

WEDNESDAY, SEPTEMBER 1, 1965

9:30 A.M. Subject: *Metropolitan Planning: Analysis and Models*

Papers: 1. An Econometric Model of Metropolitan Economy and Its Application
S. Ichimura, Osaka University, Japan

2. Data Collection Systems for Metropolitan Planning
L. S. Jay, East Sussex County Council, England

Discussants: *Z. Czerwinski*, Poznan, Poland; *K. Jones*, Norwegian Computing Center, Oslo

---

4:00 P.M. Subject: *Metropolitan Planning: Case Studies*

Papers: 1. The Development of the Metropolitan Region of Warsaw
M. Kaczorowski, Institute of Housing Construction, Warsaw

2. Greater Oslo: Considerations for Planned Expansion into the Polynuclear City Region
T. Rasmussen, Geographic Institute of Oslo University, Norway

Discussants: *A. G. Buyst*, Seminar for Applied Economics of Ghent, Belgium; *A. Sallez*, Institute for Land Use Planning and Urbanism of the Paris Region, France

---

6:30 P.M. Business Meeting

THURSDAY, SEPTEMBER 2, 1965

9:30 A.M. Subject: *New Developments in Regional Theory and Spatial Analysis*

Papers: 1. The Tensorial Approach to the Theory of Location and Regionalization
K. Dziewonski, Institute of Geography, P.A.N., Warsaw

2. Aspects of the Spatial Structure of Social Communication and the Diffusion of Information
T. Hägerstrand, University of Lund, Sweden

2:30 P.M. Subject: *Criteria for Regional Development and Investment Programs*

Paper: 1. Benefit-Cost Analysis
N. Litchfield, London, England

Discussant: *M. Zboril*, Academy of Construction and Architecture, Brno, Czechoslovakia

Summary Paper on Main Program: *W. Isard*, University of Pennsylvania, U.S.A.

FRIDAY, SEPTEMBER 3, 1965

10:00 A.M. Subject: *Systems of World Regions: Characteristics, Structure and Function*

Jointly Sponsored with the *Peace Research Society* (International)

Chairman: *M. Kaczorowski*, Polish Association for the United Nations, Warsaw

Paper: 1. Factor Analysis: A Method to Delineate World Regions
M. Megee, Missouri, U.S.A.

2. Cohesion Indices for Regional Determination
A. Szalai, Hungarian Academy of Sciences, Budapest (a communication)

Discussants: *T. B. Lahiri*, Calcutta Metropolitan Planning Organization, India; *P. Marescaux*, Faculty of Law and Economic Science, Paris

---

A Peace Research Society Session of Interest to Regional Scientists is:

2:30 P.M. Subject: *The Economics of Military Expenditures and Disarmament*

Chairman: *I. Glagolev*, Academy of Sciences of the U.S.S.R., Moscow (invited)

Paper: 1. Local Impact of Disarmament, Foreign Aid Programs, and Development of Poor World Regions: A Critique of the Leontief and Other Growth Models
M. Chatterji, University of Pennsylvania, U.S.A.

Discussants: *Th. C. Pontzen*, International Union of Peace Societies, Paris; *J. Dobretsberger*, University of Graz, Austria; *J. Dyckman*, University of California at Berkeley, U.S.A.

Other Papers of Interest to be Given at the Peace Research Society Meetings

1. Practical Applications of Game Theoretical Approaches to Arms Reduction (and to Goal Determinations among Regional Planning Authorities)
   W. Isard, University of Pennsylvania, U.S.A.
   (to be given Saturday, September 4, 9:30 AM)

2. World Population Potential Analysis and International Crises
   W. Warntz, American Geographical Society, New York, U.S.A.
   (to be given Saturday, September 4, 2:30 PM)

The September 1965 newsletter reminded members of the Twelfth U.S. Annual Meetings, November 12–14, Philadelphia and that there will be Early-Bird Ph.D. thesis Sessions in order to encourage active participation by young scholars. Graduate Students are welcome at all sessions of the conference.

It also noted that *Volume I, Papers and Proceedings of the First Far East Conference* will contain the following papers:

1. Spatial Organization and Regional Planning: Some Hypotheses for Empirical Testing by *Walter Isard*
2. Morphology and Economic Theory of the Industrial Agglomeration by *Dyodi Esawa*
3. A Model of Regional Planning by *Shinichi Ichimura*
4. A Microscopic Theory of Traffic Assignment by *Noboru Sakashita*
5. Model Plan of a New Industrial Area Through Interactivity Matrix by *Eiji Kometani*
6. The Housing Stock as a Resource by *Wallace F. Smith*
7. The "Normal" Requirements Method for Analyzing Future Employment Structures—with Special Reference to Australian Towns by *C. J. R. Linge*
8. Development Strategy for Northern and Western Alaska by *Hubert J. Gellert*
9. Industrialization Beyond the Metropolis: Current Developments in India by *J. E. Stepanek*
10. Regional Development and Employment in the Philippines by *Angel Q. Yoingco*
11. Regional Development Policy in Japan by *Saburo Ohkita*

12. Some Characteristics of Community and Regional Planning in Japan by *Eichi Isomura*
13. Some Aspects of Urbanization and Metropolitanization in Japan by *Shinzo Kiuchi*
14. Regional Allocation Criteria by *Thomas A. Reiner*

The cost per volume to members of RSA was set at $3.50.

The *November 1965* newsletter recorded that the *1966 European Meetings of the Regional Science Association* are scheduled to take place *August 29–September 1, 1966 at the Neuen Institutsgebäude, University of Vienna.* The host for the conference will be the Österreichisches Institut fur Raumplanung, Reichstratsstrasse 17, Vienna, under the direction of Professors Werner Jager and Fritz Kastner; Mr. Richard Gissner at the Institute has kindly consented to serve as local arrangements secretary. It is anticipated that this Congress will have the participation of scholars from all Continents.

Report on Fifth European Regional Science Congress, Cracow, Poland, August 30 – September 3, 1965

Those attending the Congress were greeted by Professor S. Leszczycki, President, Committee for Space Economy and Regional Planning , Polish Academy of Sciences, which served as the host of the Congress. Also present were Dr. A. Kuklinski, Secretary of the Committee, Dr. Kruczala, Local Arrangements Chairman, and several members of the Honorary Committee of the Cracow Conference.

The papers presented may be categorized as representing the *Supply* and *Demand* for new techniques. Prominent on the "supply" side were: (1)Dziewonski's tensorial approach to location theory which provides a general framework for analyzing both the forces which tend to shift, and those which tend to stabilize the location of economic activities, (2) Hägerstrand's new efforts to develop statistical estimation and simulation procedures for analyzing spatial diffusion processes, (3) Ichimura's integration of econometric and migration analyses for the projection of regional economies, (4) Klaassen's development of new indices for determining the regional economic potential of new industries for purposes of long-range planning, (5) Jay's development of new data collection, coding, and computer processing techniques for city and regional planning, and (6) Litchfield's detailed comprehensive scheme for systematically recording benefits and costs, supplemented by Burns' approach to benefit-cost estimation in a concrete situation.

The "demand" side was represented by (1) Kaczorowski's paper emphasizing the need to consider both social and psychological factors in developing more effective cost-benefit analyses, (2) Rasmussen's paper and subsequent discussion pointing up the need for extending location theory from individual plants and complexes to considerations of optimal industrial specializations for localized economic regions, (3) Bodnar's outline of problems that are and will be confronted in the various stages of integration of national economies, (4) Oishi's presentation and the subsequent discussion of the unique aspects of regional development confronting the Japanese economy and the need to attack them in a fresh manner, (5) Kuklinski's response to several papers in which he expressed the need for the development of consistent regional data banks which can be used both for regional and national planning purposes, and (6) Porwit's original synthesis of technical problems that emerge in national economic planning. The meetings were attended by nearly 300 persons from some 23 countries.

In retrospect, the Cracow conference represented the effective invasion of the Soviet bloc by regional science. Hitherto, one or a few scholars from one or more nations in the bloc attended European conferences. In contrast, at the Cracow conference there was large-scale attendance from several major bloc nations. And in subsequent years European conferences were again to take place in Cracow and Budapest, effectively incorporating into the regional science fold Soviet bloc scholars inclusive of the best at Akademgorodok, Siberia.

Report on the Second Far East Conference, Tokyo, Japan, September 15–17, 1965

The Second Far East Conference of the R.S.A. was held at the University of Tokyo, Toshi Center. Professor G. Konno, President of the Japanese Section of the R.S.A. gave the welcoming address. The following papers were presented:

1. Spatial Competition and Distance Preference; *D. Esawa*
2. Private and Public Requirements for Neighborhood Rehabilitation through Code Enforcement; *F. E. Case*
3. Urban Systems Analysis and Information Systems; *W. Z. Hirsch*
4. A Practical Application of Game Theoretical Approaches to the Resolution of Conflicts among Regional Planning Authorities; *W. Isard*
5. Economic Growth and Interregional Cooperation in Asia; *H. Arakawa*
6. On the Interregional Income Differentials in Japan: A Comment on Planning Industrial Location; *H. Nishioka*
7. Survey of the Distribution of Tangible Fixed Assets of Manufacturing Industry; *K. Baba and S. Nukaya*
8. An Econometric Approach to Annual Forecast on Regional Economy by Local Government; *Y. Kaneko*
9. Spatial Stochastic Process Models: A Method of Analyzing Spatial Point Phenomena; *T. Smith and P. Rydell*
10. Need of Comprehensive Space Management and Evolution of Agrarian Physical Planning; *T. Kano*

11. Limits of Agglomeration and Decentralization in Regional Development; *K. Murata*
12. Regional Planning in India: Problems and Prospects; *L. S. Bhat*
13. A Systems Approach to Transportation Analysis; *J. J. Cirrito*

The chairmen of the various sessions were: W. Isard, S. Okita, E. Kometani, W. Z. Hirsch, F. E. Case, and E. Isomura. The following persons served as discussants: T. Sasada, Z. Ito, T. Katagiri, M. Nakamura, N. Yokeno, J. F. Rose, T. Smith, A. Sagrista, T. Aoki, K. Susuki, F. Ueno, J. Sakamoto, H. J. Gellert, and H. Ogawa.

The papers and discussions at the Second Far East Conference will be published as Volume II, Far East Papers and Proceedings of the Regional Science Association.

It should also be noted that in 1965 the United States Congress set up the *Economic Development Administration*. Under the leadership of Benjamin Chinitz large amounts of funds for research on regional policies and social welfare were granted to several universities and nonprofit research institutes. The outcome was healthy and intensified research by regional scientists on policy issues.

The newsletters of 1966 and 1967, except for a few developments, largely dealt with routine matters of a healthy and successful world organization undergoing normal development. In the March 1966 letter there were items on: dues payments and publications; the thirteenth U.S. annual meetings of RSA, November 4–6, St. Louis (Clayton), Missouri; the Sixth European Congress of RSA, August 29–September 1 at the Neuen Institutsgebäude, University of Vienna; the Regional Science curriculum of UCLA; election of new RSA officers; extensive details on the annual meeting of the Western Section of RSA, January 28–29, 1966 at Santa Barbara, California; details on the annual meeting of the Southeastern section of RSA on April 1–2, at College Park, Maryland; the fall meeting of the New England Section of RSA, December 4, 1965 in Cambridge, Massachusetts; the first of a series of lectures on October 13, 1965 sponsored by the Regional Science Graduate Student Organization at M.I.T. and Harvard; the Council approval of the formation of a Ghana section of the RSA; Minutes of the Business Meeting on November 13, 1965 where normal operations of a business meeting are covered, namely adoption of the minutes of the previous year's meeting, a Treasurer's report, Reports of the Nominating Committee, of the Outcome of Elections, of the Publications Committee, of the Data Committee, a Notice on 1966 annual U.S. meeting of the RSA, and a long Report by Isard on Foreign Sections (French language section, Japanese section, Scandinavian section, and section organizing activity at the British Isles, Greece, Hungary, Germany, India, Ghana, and the Philippine Islands.

Clearly, the numerous and diverse activities ongoing in years 1965 and 1966 were starting to put strains on the relatively simple organization of conferences and management of the Regional Science Association.

One important non-routine development, namely the formation of the German-speaking section of RSA, took place during 1966. As reported:

Association for Regional Research, Registered Society, German-Speaking Section of the Regional Science Association (in process of formation)

The first regular meeting of the Association for Regional Research, Registered Society, German-speaking Section of the Regional Science Association (in process of formation) was held on December 11, 1965 in the Alfred-Weber-Institut, Heidelberg, West Germany. A constitution was adopted which was submitted to and later approved by the Council of the RSA. Professor Edwin von Böventer of the Alfred-Weber-Institut and many other scholars have been instrumental in forming the Section. The membership sponsoring the constitution represented a rich variety of scholarly and professional backgrounds from the German Democratic Republic, the Federal Republic of Germany, Austria and other German-speaking areas.

Except for one announcement, the remaining newsletters of 1966 concerned relatively standard matters. Those not already mentioned were: officers of the RSA sections in the U.S. and their numbers of members; a regional science research project newly established at Syracuse University; more details on the regional science curriculum at the University of California at Los Angeles; details of programs of the sixth European RSA conference; details of programs of the thirteenth annual RSA conference in North America; National Science Foundation recognition and support of regional science undergraduate research; a regional science/landscape architecture project at Harvard University; a list of officers of the association by academic year, 1957 to 1966 (updated to 1967 which appears as Appendix F); and a list of attendees by country at the Sixth European Conference: Belgium 1, Czechoslovakia 2, Denmark 5, Finland 1, France 4, Germany (FRG) 30, Germany (GDR) 3, Great Britain 3, Greece 1, Hungary 4, Ireland 1, Israel 1, Italy 8, Japan 4, Luxembourg 1, Netherlands 3, Poland 4, Portugal 1, Sweden 12, United Arab Republic 1, U.S.A. 12, USSR 6, Yugoslavia 2, Austria 43.

The one major development, which was reported on in the June 1966 newsletter was:

Second Latin American Regional Science Association Congress, Rio de Janeiro, Brazil, August 17, 18 and 19, 1966

Plans have now been completed for the Second Latin American Congress, which will be held in Rio de Janeiro under the auspices of the Ministry of Planning (Rio) and the Interstate Commission on the Parana-Uruguay River Basin (São Paulo). The meetings, scheduled to run for three days August 17–19, will also serve to conclude a training Institute in Regional Science and Regional Development sponsored by these Brazilian agencies. As with all Regional Science Association meetings, all interested persons are cordially invited to attend the Congress.

Among the contributions which have been scheduled for these meetings are papers by Mario Brodersohn, di Tella Institute, Buenos Aires (Testing Interregional Input-Output Assumptions: A Case Study of Argentina); Morris Hill, University of

North Carolina (Uses of Cost Benefit Analysis in Regional Development); Walter Isard, University of Pennsylvania (Problems of Cooperative Development for the Regions of a System); Laveen Kanal, Philco Corporation (Pattern Analysis and Related Spatial Statistical Techniques); Anthony Pascal, U.S. Economic Development Administration (The Regional Development Institute of the Economic Development Administration, U.S.A.); Richard Quandt and William Baumol, Princeton University (The Abstract Transportation Mode Model: Theory and Measurement); Carl Peter Rydell, University of the City of New York (Regional Growth Models); Hans Singer and Salvatore Schiavo Campo, U.N. Economic and Social Council (Wages, Skills and Localization of Industry); Benjamin Stevens, Regional Science Research Institute and University of Pennsylvania (Programming New Industry); and Jose Villamil, University of Pennsylvania and University of Puerto Rico (Development Problems). In addition, significant contributions are expected to be made by the cooperating Brazilian organizations. As with the first Latin American Regional Science Association Congress, held in the fall of 1962 in Caracas, Venezuela, with the cooperation of the University of Venezuela, CENDES, this should prove to be a most stimulating set of meetings focusing on important problems of regions undergoing development,

Those interested in obtaining further information about the Congress and about local arrangements (including accommodations) should write to the recently appointed Administrative Secretary for the Congress: Sr. Luis Rocha Neto, Escritorio de Pesquisas Economica Aplicada, Ministerio de Planejamento, Palacio da Fazenda, 80, Rio de Janeiro -GB, Brazil.

The newsletters of 1967 provided information on: dues payments and publications, elections, the fourteenth U.S. annual meetings, November 3–5, 1967, Harvard University and its tentative programming; additional details on other conferences already noted; the Southeastern, Western and New England sections meetings, and a report on the second Latin American congress; the forthcoming third Far East RSA conference September 12–14, Tokyo; the forthcoming London meetings of the newly formed British section, August 26–27, 1967 and some papers tentatively scheduled; the formation of the Brazilian section of the RSA, formally approved by the Council; and active interest in the formation of an Argentine section of RSA.

There were also the announcements of the following units:

Center of Regional Science at the Universite d'Aix-Marseille, France

A Center of Regional Science (Centre de Science Regionale) has recently been established at the Universite d'Aix Marseille, Marseille, France. The Center consists of a federation of four participant faculties from the University: (1) the Center of Regional Administration Science of the School of Law and Economics; (2) the Center for Research on Regional Economic Development of the Institute for

Business Administration; (3) the Center of Urban Sociology of the School of Humanities; and (4) the Section of Regional Geography and Spatial Organization of the Geography Department in the School of Humanities. This Center's main purpose is to undertake multidisciplinary research on regional problems in France. It is currently engaged in a study of regional economic and social development as it is influenced by decision making at the national level.

Institute for Transportation, Tourism and Regional Science at the Copenhagen School of Economics and Business Administration, Copenhagen, Denmark

An Institute for Transportation, Tourism and Regional Science has been established at the Copenhagen School of Economics and Business Administration. The Institute presently offers a program of courses which constitute a field of specialization for master's degree candidates in the business administration program. In introducing this new field of study into the curriculum, the Copenhagen School of Economics and Business seeks to give graduate level preparation to students looking for careers in management, the tourist industry, and in public administration for town, regional, and national planning. In addition, a major research project of the Institute is the publication of a five volume series dealing with the international aspects of marketing, tourist travel, international congresses, industrial location, and urban development. The Director of the Institute is Professor Ejler Akljaer.

Also, there were two major developments in 1967. The newsletter announcement of one was:

Regional Science Institute Founded at the Technische Hochschule Karlsruhe, Karlsruhe, Germany

A Regional Science Institute (Institut für Regionalwissenschaften) has been founded at the Technische Hochschule Karlsruhe, Karlsruhe, Germany. The Institute is presently composed of ten scholars representing the disciplines of economics, architecture, geodesy, geography, city and regional planning, sociology, and transport engineering. Major emphasis will be on the teaching of regional science and regional science research projects.

This Institute later initiated and encouraged the holding of annual summer research training institutes of several weeks' duration in Europe. The Institute has become an ongoing activity. It has played a significant role in instructing scholars on the new regional science methods—scholars mostly from Europe, but also from other parts of the world.

The second major happening was the initiation of regional science activity in India. This took considerable effort on the part of Isard, Chatterji and Reiner but in time it led to the formation of the Regional Science Association of India, and the publication of the *Indian Regional Science Journal*, an undertaking that had to be subsidized by the North American office for several years. This development represented another invasion of regional science in a major area of the world. Thus one more large segment of the world's population was brought into the regional science fold. Unfortunately, because of the lack of hard currency, Indian scholars were unable to participate in as effective a manner as was hoped with those from Europe, Japan and North America.

One interesting speculation of the late 1970s and early 1980s concerned the possibility of annually scheduling conferences and arranging with airlines round-the-world trips. One trip, for example, might start along a leg from a North American conference location (say New York, Boston or Philadelphia to which Japanese, North American scholars and perhaps a few European scholars might travel) to a European conference site (say, London, Amsterdam or Frankfurt), then along another leg to an Indian conference site (New Delhi, Calcutta or Bombay, at which they would be joined by Indian scholars), and then along a third leg to a Japanese conference site (Tokyo, Osaka or Tsukuba), and finally a return to North America to complete the conference circle. Those European scholars who did not start the conference circle at the North American location might end up with a trip to the Department of Regional Science in Philadelphia (where seminars and lectures on advanced regional science models might be arranged to comprise the third leg of their conference circle). Scholars from India (only a few of whom were expected to be able to afford a world trip) could join the World Trip club at an Indian conference site, could also participate in the Japan conference and then the above seminars and lectures in Philadelphia, and next cross the Atlantic Ocean to engage in scholarly exchanges at one or more leading European universities. This speculative round-the-world travel was never realized but it testifies to the fact that the RSAI had become a major global institution, with rather intense interaction among many widely separated global parts. This interaction was already much more intense than that among the geographers of the International Geographic Union or among economists of the International Economic Association. The RSAI had become a truly leading international scholarly body.

With the above speculative scheme and reports, it is appropriate to end this Part I of the History of Regional Science and the Regional Science Association. Clearly, a set of basic organizational changes had to be instituted were the rapid globalization of the Association and its scholarly advancements to continue. Such were gradually initiated and effectively accomplished during the next period by David E. Boyce, who first joined Isard in global activity at the 1968 Budapest conference. Incidentally, this conference took place at the heels of the Soviet invasion of Czechoslovakia, a horrific event. The successful and uninterrupted program of the Budapest conference testified to the strength of the global bonds already established among the non-Soviet and Soviet (European and Siberian) scholars. There was little, if any, negative effect of the invasion upon these scholarly bonds, then and later.

8 My Current Thinking on the Scope and Nature of Regional Science and Opportunities for Its Advance in Basic Research and Policy Analyses

With my part of the history of the Regional Science Association, International now complete, I would like to present some history of my thinking about and approaches to location, regional and spatial research.

My early thinking on location theory and the closely related spatial analysis has essentially been covered in the material presented in the Sections 1, 2 and 3 of this book dealing with the pre–Regional Science Association era. In 1954 when the Association had been established and during the hot debate about regional science in the immediate years thereafter, much of my broader thinking about the scope and nature of regional science took form.[22] Today, I still view regional science as a social science field, focusing upon spaces and systems of spaces, regions and systems of regions, location and systems of locations. Distances—physical, economic, political, social and cultural—are key concepts. Basic phenomena embrace, among others, transportation and communications, population and commodity flows, and land, water and other resource use and in general development and wise environmental conservation, management and control. Some significant characteristics of concern are population numbers and its distribution, income and its distribution among classes, employment (total and by type), diverse metropolitan and rural patterns, administrative area and hierarchical structures. Change—growth and decline—and projection are of the essence of regional science; and comprehensive interdependence and linkage analysis and conflict management strongly color its approaches, both behavioral and mechanistic.

Regional science has important interdisciplinary aspects (to be noted below). As a new field of investigation, it draws generously upon the concepts, methods and techniques of existing social science disciplines. But regional science has become more than a serious interdisciplinary pursuit. It has come to have its own peculiar core and wholeness which is beyond the mere addition of regional elements of the diverse existing social sciences; and it is forging tools and techniques accordingly. In one sense it may be viewed as treating societal units having spatial dimensions

[22] The statements that follow are modifications of statements in my 1960 article on the scope and nature of regional science.

and possessing certain mysteries of structure and function which established disciplines are unlikely to discover. From another perspective, it is more than Alonso's narrower statement that regional science can possibly be viewed as a set of points relating to subject matter, the intent and approach of which contains at least one subset not shared with any other element of that larger, comprehensive set which may be designated, all disciplines.

Let me elaborate further some of these controversial points. First, in terms of past practice, current interests of those associated with the several regional science units, and likely future developments, regional science is primarily social science. It is concerned with the study of man and the spatial forms which his continuous interaction with, and adaptation to, physical environment take. Regional science concentrates its attention upon human behavior and institutions; and, unlike geography, is much more confined to advanced scientific analysis of social processes, giving much less attention to spatial detail and associated physical and biological elements.

Or regional science may be viewed as a field of inquiry which, in the words of the sociologist Whitney (1957, p. 27), presses upon the task of developing interregional frameworks which express 'all kinds of linkages in proper relation to one another, that is, as a complete and self-sustaining closed system.' This second point can be better developed if we quote at length from Whitney.

Sociologists recognize that there are areas of varying size which are different from other areas not simply in their physical characteristics, but also in terms of their aggregate social organization … 'Region' is not a mere synonym for 'area', of course. A regional entity cannot fall below a minimum size determined by the smallest area which cannot support a distinctive social organization. It cannot be larger than the maximum size within which such separate social organization can be maintained … Every region will display distinctive functions; and these functions necessarily derive most directly from the characteristic social and cultural organization and indirectly reflect the limiting features and the positive possibilities inherent in the physical base …

In a modernized society, we may assume a fair to high degree of mobility. It follows that the relation of regions to one another in such a society is one of interdependence, showing the organic solidarity described by Durkheim and based on the complementary activity of specialized parts. Not only do we have interdependence, but a high degree of interdependence since this is a requisite for developing and maintaining a modernized economy and society …

With respect to the specific identifications of bonds which link regions and make possible their functional specializations, sociologists have been concerned with the communication process, communication mechanisms, mobility and migration … They have not, however, attempted to develop a general construct which would include all interregional bonds, which would measure with some degree of accuracy the volume of interregional exchange of all kinds ... Nor are they properly qualified nor called upon to undertake such an all-inclusive task ... Sociology as a discipline must be restricted to its own specialized approach. Thus, like every other social science, it can contribute to, but it cannot single-handedly develop, an interregional model which expresses all kinds of linkages in proper relation to one another, that is, as a complete and self-sustaining closed system ... This is a task which may legitimately be attempted by regional science. It may never be accomplished, but it is a proper goal ... (Whitney, 1957, pp. 26–27).

Although at the time Whitney made these statements my view of a region was somewhat different than his, and although I made more explicit the definite limits to the field of regional science and the high probability that it will fall far short of developing an interregional model inclusive of all interregional linkages, I was stimulated by his last statement. In accord with it, I did feel in the 1950s that it was important to embrace the approaches of political science, history, anthropology, geography, planning, as well as sociology to capture in an interregional model all kinds of linkages in proper relation to one another. I made efforts to do so in joint sessions with the American Sociological Society, the American Political Science Association, the American Institute of Planners, the American Association of Geographers, the American Association for the Advancement of Science, and I did distribute to regional scientists seminal papers by scholars of political science, sociology, planning, geography and history. But by the end of the 1950s no significant contact with political scientists, sociologists, historians and anthropologists had been achieved. Significant interaction, however, was established with geographers and planners. As a result, my thinking with regard to regional analysis took on a more integrative character going well beyond the simple addition of elements of geographic and planning analysis to those of economics in which I was trained. It culminated in the 1960 *Methods of Regional Analysis* book which embodied fusion of analytic approaches already discussed.

Nonetheless, in the 1960s I was interested in *theoretically* capturing political and social forces within an interregional framework. I started with a statement on the General Equilibrium of the Economic Subsystem in a Multiregional Setting (Isard and Ostroff, 1960). To add more of reality to this statement of the multiregional system, I undertook a more ambitious and comprehensive project which resulted in the 1969 book on *General Theory: Social, Political, Economic and Regional* and related articles. This attempt to extend the interregional equilibrium framework to embody the political and social subsystems also recognized the importance of another key area of analysis—namely, the analysis of the interaction of decision makers (individuals, organizations and institutions) and their interdependent decision making in situations of conflict over policy and other joint actions. This led to attempts at formal treatment of such in policy spaces involving two or more parties and coalition possibilities.

It was not long before it became necessary to introduce dynamics into regional research and accordingly the analysis of dynamical systems. Thus emerged relevant articles and my 1979 book (with Liossatos) on *Spatial Dynamics and Optimum Space-Time Development*. Here, given the successful applications of the gravity model, we explored the possibility of finding from physics, chemistry, biology and the natural sciences as a whole, other concepts and analysis that could find gainful use in the study of regional systems and their subsystems. Field description of spatial interaction, a two equations of motion framework and distributed dynamical systems were looked into. Some formalization of hierarchical theory was attempted. Bifurcation and Thom catastrophe theories were employed to model structural change, often sharply discontinuous, that can be observed in the development of a region, and also at times in the sequence of transitions in other regions of an interregional system. Into this type of study the use of the *Master Equation* was investigated to embed a deterministic equation of motion in a stochastic process. Lastly, speculations were set forth on a geometric description of spatial interactions stemming from general relatively theory.[23] Concomitant with the above research of the 1970s and 1980s there was much investigation of the ecologic system and environmental problems, on local, regional and world scales. (See *Ecologic-Economic Analysis for Regional Development*, 1972.) And by the late 1980s and early 1990s the environmental issues became increasingly embodied along with trade and regional development ones in conceptual Global Models (also viewed at times as Linked Integrated Multiregional Models).

Subsequent to the construction of Global Models, and to the advanced theoretical ones of the 1980s—advanced in terms of mathematical analysis, but admittedly unrealistic in terms of the highly questionable assumptions required to realize their acclaimed results—there was a return to the less elaborate and more applicable research approaches of earlier decades. There appeared (with colleagues) my 1998 *Methods of Interregional and Regional Analysis*. Perhaps in this book too much adherence was given to the structure of the very successful fusion of the methods of regional analysis achieved in the 1960 *Methods of Regional Science* book. Nonetheless, the 1998 book extended existing methods, for example, regional and interregional social accounting, and presented several new ones that can be fruitfully incorporated in regional research. One is microsimulation, the scope of which is currently being greatly enlarged on an interregional framework, being stimulated by advanced computer technology. Another is the combination of (1) AGIE (Applied General Interregional Equilibrium Analysis), a spatial extension of CGE (computable general equilibrium analysis), and (2) nonlinear interregional programming. Still another is the use of conflict management procedures (CMPs),

[23] Associated with relativity are concepts of space and time as *active* containers and not passive ones—of curved space-time. The exploration of such concepts for use in regional science I find highly tantalizing.

for example the use of pairwise comparisons and relative utility, to suggest "mutual improvement" joint actions (policies) for regions, organizations, individuals and other decision-making units locked in conflict.[24] Going beyond the 1998 book there are new developments in the use of agent-based models that enable a researcher to handle better the problem of gaining further understanding of what the impacts of random elements may have been in past regional and other developments and what they may be in future and planned developments—a problem fully recognized in the 1960 Methods book but impenetrable because of the lack of computer capability. Then there are other approaches—for example, the location science approach of operation researchers in attacking problems of scarce resources, preservation of unique ones, and in overall management; here more extensive use of integer programming in combination with other methods can be involved. Then there is current exploration of the increasing complications of evolving "complex systems" that characterize the creative work of Hewings, Sonis and their associates and of others which perhaps may revolutionize regional science methods.

The above development of my thinking and my work on research methods and that of others were accompanied over the last fifty years by major advances in understanding of the problems and structure of both the global system and all its various and specific parts. These parts are:

a. its major, extensive regions (continental, supranational and other blocs of nations, some new) and huge aggregates of regions (such as Siberia);
b. its very nations, each being a political region;
c. China, its resource and metropolitan regions and their hierarchical and nonhierarchical units of diverse sizes and characteristics;
d. its subnational regions (groupings of states and provinces), and each state and province itself; and
e. its variegated localities (towns, villages, and still smaller rural units).

In contrast, at the birth of regional science, the global system was relatively simple. At that time, too the global system as a whole was not of deep concern to regional scientists. In our eyes, except for international economic trade, it was largely a

[24] It should be noted that oftentimes an effective conflict management procedure for a given conflict requires *art* as well as *scientific analysis*. This is so since effective mediation requires obtaining from parties involved in a conflict their perceptions of relevant variables, the relative weights they apply not being subject to what is commonly regarded as scientific analysis. See Isard and Chung, 2001. But perhaps this may turn out to be one way of quantifying certain political and sociological variables in ways that will allow us to fuse them with current economic and regional science factors.

political system, for analyses of which balance of power and the like power-oriented concepts of political science dominated. And since trade between nations was largely influenced by the intricacies and fluctuations of foreign exchange rates that governed world prices, we location theorists and regional development analysts judged it best not to be involved with such. Specifically, we left the analysis of industry location and economic development based on trade between nations to economists specializing in trade theory and policy. In effect, we concentrated on industrial location and development analysis *within* the regions of any given nation being investigated. For example, see my chapter 9 on Some Basic Interrelations of Location and Trade Theory in Isard (1956).

Fifty years later, the situation is different. First, we have observed and still observe that the research projections (forecasting) of foreign exchange movements by specialist economists are much to be desired (if not impossible). This is largely so because of (1) random events (reflecting the need to employ to some extent agent-based models in international trade theory), (2) the need to be more effective in embracing noneconomic factors (political, sociological, cultural) in their models, and (3) the need fully to employ interregional analysis. (It can be said that some economists—the regional economist type—have embraced to some extent interregional analysis, but still most inadequately.) Moreover, the pursuit of standard economic doctrine—unduly affected directly or indirectly by the highly abstract mathematical models of economists—has led to a number of miserable failures in policy-making by the International Monetary Fund and the World Bank.[25] Much more consideration of noneconomic factors along with interregional analysis fully sensitive to spatial relations is yet to be satisfactorily embodied in their models of development. While in recent years regional scientists have already moved somewhat in this direction, much more needs to be done. An interregional policy based on some combination, desirably fusion, of efficiency and equity principles must be achieved. Such is now beginning to emerge in problems being attacked by the new European Union—an agglomerate of both nations and regions of nations where the sovereignty of nations is downplayed. There, interdependence and linkage analysis of the various regions and nations is being brought to the fore. However, on the larger global scale, and in particular on the problem of better combined (fused) efficiency and equity principles for urgent development of the many poor regions of the world outside the North American/European orbit, little is being

[25] In particular, I have in mind the "accepted" economic principles of financial management that have been embodied in the conditions set forth for financial assistance by the International Monetary Fund. These conditions reflecting the blindness of the Fund to cultural-political structure have led to disastrous outcomes, for example, in Indonesia in the 1990s and in Argentina in the early years of the twenty-first century.

accomplished. Here there are enormous opportunities[26] for involvement by regional scientists with their interregional (linkage) analysis when combined with their greater inclination to employ a multidisciplinary framework.

Leaving the more aggregate patterns of the multinational corporations and organizations, associated business leaders, other supranational and national actors in the global system, consider its smaller parts, components and groups and less prominent leaders and decision makers. At these levels, there have occurred major changes over the last fifty years. There has been, as Markusen (2002) points out, much devolution of responsibilities and taxing powers toward lower levels of government, not only in the older major industrialized countries but also in the newer developed and less developed ones. Here, from their close association with urban and regional planners, regional scientists have had excellent opportunities to be exposed to the need for a cross-disciplinary approach in their research. To a modest extent they have incorporated such in their analyses.[27] They, nonetheless, must involve more inputs from other social sciences, especially political science, sociology and psychology. Such is essential for developing more effective research on structure and problems of metropolitan regions.

I close by maintaining that there are vast vistas for regional science research on our current global system—ranging from the most aggregate meaningful composites of spaces and areas down to the smallest, meaningful spatial (areal) units. With our latest developments in interregional analysis we are able to go well beyond what economists can do (we already have caught up to them in understanding and even proliferating advanced quantitative and mathematical analyses, many admittedly based on unrealistic assumptions). With our interregional and linkage approach and our sensitivity to the importance of key administrative, cultural, political and other social factors regional scientists can also achieve much more realistic and effective studies.

---

[26] The existence of these opportunities for regional scientists reflects the fact that via the Regional Science Association, International, regional scientists have formed a *highly interactive global (or international) club*. On average each member has greater appreciation, knowledge and greater willingness to give weight to the different perspectives of cultures (each different from his own) than scholars in other more specialized social sciences.

[27] This was so especially in the early years of the Regional Science Association when regional scientists were prominent among the leaders in urban research.

Appendix A

Memorandum on a Census Monograph on the Location of Economic Activity and Its Relation to Population

I

The purpose of this memorandum is to present suggestions on the topics and the organization of a census monograph on the location of economic activity and its relation to population. The monograph would utilize data from the 1947 Census of Manufactures, the 1948 Census of Business, the 1950 Census of Population and Housing, and from previous censuses.

In line with established census divisions, the broad term economic activity may be divided into agriculture, mining, manufacturing, and business activity. However we shall not discuss the location patterns of agriculture and mining for we feel that personnel in the Department of Agriculture and the Bureau of Mines are more qualified to do so and more familiar with existing data and current special studies.

With reference to the location of manufacturing and its relation to population, the proposed monograph should be a continuation of previous census monographs:

1. F. S. Hall, *The Localization of Industries,* Twelfth Census of the United States: 1900; Volume VII Manufactures, Part I; pp. cxc–ccxiv.
2. U. S. Bureau of Census, Location of Manufactures, 1899–1929, A Study of the Tendencies Toward Concentration and Toward Dispersion of Manufactures in the United States, Washington, Government Printing Office, 1933.
3. U. S. Bureau of Census and Bureau of Agricultural Economics, Changes in the Distribution of Manufacturing Wage Earners, 1899–1929, Washington, Government Printing Office, 1942.

Part II of this memorandum contains the details of our suggestions on this subject.

Preceding census monographs have not covered the location of business (service or tertiary) activities and its relation to population. With increasing attention being paid to such activities, in terms of measuring economic welfare, and in terms of new employment opportunities, a study of their location patterns seems highly desirable. If it is not undertaken elsewhere, it should be included in this monograph, along with the study of the location of manufacturing, in the manner outlined in Part III.

Part IV discusses the geographic patterns of the major and minor occupations groups, and regional and state specialization. Also, there are suggested additional compilations and cross tabulations that would be useful to both economists and regional planners. If these items are not covered in other projected census monographs, they should be included in this one.

II

The study in the projected monograph of the location of manufacturing in its relation to population should not only cover the existing geographic patterns of manufacturing and population, but also the locational shifts that have been taking place since the first Census of Manufactures in 1899. Analysis of the historical shifts in the location of population and manufacturing and the contrasts between these shifts is of particular value in projecting and planning regional development. Therefore the proposed tables, maps, and charts present the changing historical locational patterns as well as current ones.

As we discussed in previous census monographs, data on employment and occupations are the most appropriate for detailing these shifts. The suggested tables, maps, and charts listed below arrange these data in a way which facilitates locations analysis. Many of these are a continuation of those in previous census monographs.

A. The following maps give a picture of the location of manufacturing and its changes contrasted to the location of population and its changes, by census regions, states, and counties.

1. (a) Dot map of the location of manufacturing wage jobs in the United States in 1947, 1947 Census of Manufactures.

   (b) Dot map of the location of population in the United States in 1950, 1950 Census of Population and Housing.

2. (a) Bar map showing percent change in the number of manufacturing wage jobs in the United States by census regions, for consecutive census years, 1899–1947.

   (b) Bar maps showing percent change in the population in the United States by census regions, for consecutive census years, 1900–1950.

3. (a) Shaded map showing percentage change in the manufacturing wage jobs in the United States, by counties, 1939–1947.

   (b) Shaded map showing percentage change in the population in the United States by counties, 1940–1950.

Associated with the above maps should be these tables:

1. Average number of manufacturing wage jobs by census regions and states, for each census year, 1899–1947.
2. Population by census regions and states, for each census year, 1900–1950.
3. Percent distribution of manufacturing wage jobs by census regions and states, for each census year, 1899–1947.
4. Percent distribution of population by census regions and states, for each census year, 1900–1950.
5. Percent change in manufacturing wage jobs, 1899–1947, and population, 1900–1950, by census regions and states.
6. Percent change in manufacturing wage jobs, 1899–1919, 1919–1929, 1929–1939, 1939–1947, and in population, 1900–1920, 1920–1930, 1930–1940, 1940–1950, by census regions and states.
7. Percent change in the number of manufacturing wage jobs per thousand population, 1899–1919, 1919–1929, 1929–1939, 1939–1947, by census regions and states.
8. Number and percent counted in labor force 1940 and 1950, and counted as gainful workers 14 years and over, 1900–1930, by census regions and states.
9. Average number of manufacturing wage jobs, and manufacturing wage jobs per thousand in labor force, 1950 and 1940, and per thousand gainful workers, 1900–1930, by census regions and states.

B. The following graphs and tables contrast the location pattern of manufacturing and population and their changes, by city size groups and by urban-metropolitan type areas. They are of special significance for analyzing trends in the concentration, diffusion, and dispersion of manufacturing and population.

1. (a) Line graph for manufacturing wage jobs in the United States for total United States, and by industrial areas, and outside industrial areas, for census years during the period, 1899–1947.

   (b) Line graph for population in the United States for total United States, and by industrial areas, and by outside industrial areas, for census years during the period, 1900–1950.

2. (a) Line graph for manufacturing wage jobs in the United States for total United States and by city size groups based on 1930 (perhaps 1940) census, for census years during the period, 1899–1947.

   (b) Line graph for population in the United States for the total United States and by city size groups based on 1930 (perhaps 1940) census, for census years during the period, 1900–1950.

3. (a) Line graph for manufacturing wage jobs in the United States for total United States and by urban-metropolitan type areas, for census years during the period, 1899–1947, or to the extent to which materials are available. For this purpose Creamer[28] has used the following urban metropolitan type areas:

    1. Principal cities in the census industrial areas
    2. Satellite cities in the census industrial areas
    3. Remainder of the industrial area
    4. A city of 100,000 or more population not included in an industrial area
    5. Remainder of the county in which (4) is located
    6. Important industrial counties with no city as large as 100,000 population
    7. All other United States

    However, the categories used should be the ones employed in other census monographs on urban-metropolitan type areas, which would not necessarily be the ones above.

    (b) Line graph for population in the United States for total United States and by urban-metropolitan type areas, for census years during the period 1900–1950, or to the extent to which materials are available.

Associated with the above graphs should be these tables:

4. Average number and percentage distribution of manufacturing wage jobs and population, wage jobs per thousand population and percent increase in the United States for total United States, industrial areas, and outside industrial areas, for the census years during the periods 1899–1947 and 1900–1950 respectively.
5. Average number and percentage distribution of manufacturing wage jobs and population in the United States and in census regions by city size groups based on 1930 (perhaps 1940) census, for the census years during the periods 1899–1947, and 1900–1950, respectively.
6. Average number and percentage distribution of manufacturing wage jobs and population of United States and of census regions by urban-metropolitan type areas, for the census years during the periods 1899–1947, and 1900–1950, respectively.

---

[28] Daniel B. Creamer, *Is Industry Decentralizing?*, Philadelphia, University of Pennsylvania Press, 1935, pp. 5–6.

C. It is also desirable to present maps, charts, and tables similar to those detailed above in sections A and B for major industrial groups and selected industries. There would be some modifications in constructing these for industrial groups and individual industries, but because of space limitations, we have not listed these changes.

The major industrial groups might be those used by the census, or those used by Leontief and the Interindustry Division of the Bureau of Labor Statistics, or those used by other research bureaus. Particular industries should be selected only after consultation with interested agencies.

D. The preceding sections provide maps, tables and charts which are basic to the locational analysis of manufactures and its relation to population. However the usefulness of the monograph would be considerably increased by including certain computed coefficients, which, in a sense, summarize the data of the preceding tables. These are:

1. *The Location Quotient.* This measures the degree to which any industry is localized in a particular area relative to all manufactures. It is computed by dividing the percentage of total national wage jobs in a given industry in a given area by the percentage of the total national wage jobs in all manufacturing in the same given area. For example, the state of Michigan had:

$$\frac{\text{63.8\% of all U.S. wage jobs, in the automobile industry}}{\text{6.61\% of all U.S. wage jobs in all manufacturing}}$$

This yields a location quotient of 9.62 for the automobile industry in Michigan.[29] A location quotient of one represents a norm in which the area's percentage of wage jobs in a given industry is the same as its percentage of wage jobs in all manufacturing.

This quotient should be computed with reference to census regions and metropolitan areas for major industry groups and selected industries for 1947 and for previous censuses in order to cast light on trends in the industrial composition of regions.

2. *The Urbanization Quotient.* This measure indicates the degree of concentration of any given industry in any given urban-metropolitan type area (or city size group) in relation to all manufacturing. It is computed by dividing the percentage of total national wage jobs for a given industry in a given urban-metropolitan type area by the percentage of total national

[29] National Resources Planning Board, *Industrial Location and National Resources*, Washington, Government Printing Office, 1943, p. 107.

wage jobs in all manufacturing in the same urban-metropolitan type area. It is essentially a location quotient. A result of one indicates "that there is no difference between the proportion of employees in the given industry in that area and the proportion there engaged in all manufacturing."[30]

The urbanization quotient should be computed for major industry groups and for selected industries for 1947 and previous census years. Its change over time would indicate the changing relative degree of concentration of industry groups and particular industries in various urban-metropolitan type areas (or city size groups).

3. *The Coefficient of Localization.* This coefficient, which in certain contexts is called the coefficient of geographic association, measures the degree of concentration in the geographic pattern of any item relative to the geographic pattern of any other item. For example, in contrasting a given industry with all manufacturing, it is computed by (1) summing for all states (or regions) the plus deviations between the percentage of total wage jobs of the industry in any state and the percentage of total manufacturing wage jobs in the same state and (2) dividing by 100. A zero coefficient indicates complete correspondence, a high coefficient considerable non-correspondence.

   Where data are available, such coefficients should be computed for the following items for both current and previous censuses to show changes in the degree of geographic association.

   (a) population against area, for each census year during the period, 1900–1950

   (b) all manufacturing wage jobs against area for each census year during the period, 1899–1947

   (c) wage jobs in each two, three, and four digit census industry against wage jobs in all manufacturing, for each census year during the period, 1899–1947.

   (d) wage jobs in each two, three and four digit census industry against population, for corresponding census years during the periods, 1899–1947, and 1900–1950, respectively.

   (e) for production processes that are carried on in stages, as the manufacture of bread and steel, it is desirable to trace out the geographic linkage of the various stages by the use of this coefficient. This should be done for selected sectors of the economy, starting with the geographic pattern of the raw material and ending with the geographic pattern of population.

[30] Ibid., p. 105.

4. *The Localization Curve.* This is a less aggregative way of depicting the extent and form of the geographic relationship between a given industry and population. Drawn on a graph with percentage of population measured along the horizontal axis and percentage of total operatives of a given industry along the vertical, the curve cumulates population and occupation simultaneously, beginning with the local units with the highest ratio of operatives to population.[31] The extent of the departure from a line drawn forty-five degrees from the base line indicates the extent of localization, and the slope of the curve at any point indicates the specialization for the local areal unit represented by that point. This method has the advantage of showing both the total localization and the pattern of specialization among the local areal units.

   Localization curves might be computed for selected industries from 1900–1950, to the extent data are available. It would provide another measure of changes in the concentration and dispersion of sectors of the American economy.

III

We now turn to the study of business activities in its relation to population. Here analysis in terms of urban-metropolitan regions as well as customary census regions is desirable. Obviously the locational structure of many activities is meaningful only when linked to metropolitan cores and their tributary areas. Though many of the maps, charts, and tables detailed below could be included in section II with manufacturing, we have not done so since we do not know the scope of other projected monographs. The system of numbering in this section is a continuation of that in the previous section.

Also, since we do not know whether this projected monograph will contain business activity, we are merely listing possible comparison data, census year by census year, without inquiring into their comparability.

A. The following maps depict the location of business activity and its changes, contrasted with the location of population and its changes, by census regions and states, and/or urban-metropolitan regions and counties.

   1. (c) Dot map of the location of personnel in business activity in the United States, 1948 (contrasts with 1(b) in section II).

[31] Edgar M. Hoover, Jr., "The Measurement of Industrial Localization," *The Review of Economic Statistics,* Vol. XVIII, 1936, p. 164.

2. (c) Bar map showing percentage change in business activity personnel in the United States by urban-metropolitan regions for consecutive census years, 1929–1948.

   (d) Bar map showing percentage change in population in the United States by urban-metropolitan regions, for consecutive census years, 1930–1950.

3. (c) Shaded map showing percentage change in business activity personnel in the United States, by counties, 1938–1948 (contrasts with 3(a) above).

Associated with the above maps should be these tables:

1a. Average number of business activity personnel by states, census regions and urban-metropolitan regions for each census year, 1929–1948.

2a. Population by urban-metropolitan regions for each census year, 1930–1950.

3a. Percentage distribution of business activity personnel by states, census regions, and urban-metropolitan regions for each census year, 1929–1948.

4a. Percentage distribution of population by urban-metropolitan regions for each census year, 1930–1950.

5a. Percent change in business activity personnel, 1929–1948, and population, 1930–1950, by states, census regions, and urban-metropolitan regions.

6a. Percent change in business activity personnel, 1929–1938, 1938–1948 and in population, 1930–1940, 1940–1950, by states, census regions, and urban-metropolitan regions.

7a. Percent change in the number of business activity personnel per thousand population, 1929–1938, 1938–1948, by states, census regions, and urban-metropolitan regions.

B. The following graphs and tables contrast the location pattern of business activity and their changes, by city size groups, and by urban-metropolitan type areas. They are of special significance for analyzing trends in the concentration, diffusion and dispersion of business activity and population.

2. (c) Line graph for business activity personnel in the United States for total United States, and by city size groups based on the 1930 (perhaps 1940) census, for census years during the period, 1929–1948.

3. (c) Line graph for business activity personnel in the United States for total United States, and by urban-metropolitan type areas, for the census years during the period, 1929–1948.

Associated with the above graphs should be these tables:

2a. Average number and percentage distribution of business activity personnel and population in the United States in census regions, and in urban-metropolitan regions by city size groups based on 1930 (perhaps 1940) census for the comparable census years during the period 1929–1948, and 1930–1950, respectively.

3a. Average number and percentage distribution of business activity personnel and population of the United States, of census regions, and urban-metropolitan regions by urban-metropolitan type areas for the census years during the periods 1929–1948, and 1930–1950, respectively.

C. It is also desirable to present maps, charts and tables similar to those detailed above in sections A and B for major and minor breakdowns of business activity, and for selected individual business activities. Because of similarity, we do not detail them here. However, some of these can be in terms of sales and value of services, as well as personnel engaged.

D. The location quotient, the urbanization quotient, and the coefficient of localization can be computed to summarize data of the preceding tables. Procedures and suggestions are not listed here, since, for the most part, they are the same as in II(D).

E. 1. Line graph of the percentage distribution of

(a) business activity personnel

(b) business sales (and value of services)

(c) per capita business sales (and value of services) for central metropolitan cities, and surrounding cities by five or ten mile zones, for the United States, 1938 and 1948.

2. Line graphs similar to the above for each major and minor breakdown of business activity and for selected individual business activities.

Associated with the above line graphs should be tables containing the data used in the graphs.

IV

A. The occupational data of the Census of Population can be used to indicate for geographic areas, the division of employment by major and minor groups of occupations. This is desirable to show the broad lines of areal specialization within the United States. Detailed below are maps and tables to depict this specialization:

1. Bar maps of the percentage distribution of employed workers of each state among major groups of economic activity, 1940 and 1950. (Agriculture, mining, manufacturing, etc.)
2. Bar maps of the percentage distribution of employed workers of each census region, among major groups of economic activity, 1940 and 1950.
3. Bar maps of the percentage distribution of employed workers of each census region, among minor groups of economic activity, 1940 and 1950.

Associated with the above maps should be the following tables:

1. Average number and percentage of employed workers in each state and census region, by major and minor groups of economic activity, 1940 and 1950, and for previous census years in which data are available.
2. Percent change in employed workers in each state and census region by major and minor groups of economic activity, 1940–1950, and for previous sets of consecutive census years for which data are comparable.

B. The 1947 data on value of output (shipments) for each industry by states or census regions, in conjunction with technical production coefficients currently being computed by the Interindustry Division of the Bureau of Labor Statistics, permit estimates of the consumption of various industrial products by states or census regions. Such data can be of great value, not only to students of location, but to economists concerned with regional problems and market analysts.

For any area consumption estimates of the product of any given industry may be derived in the following fashion. The dollar output of each industry of the area is multiplied by a corresponding technical production coefficient which indicates the amount of the product of the given industry used per dollar of output of the relevant industry. Summing the resulting products yields the consumption estimate of the product of the given industry for the given area.

C. Other data that might be considered for inclusion in the projected monograph, if they are not presented elsewhere, are:

1. Income data by counties, states, and census regions if the Census of Population data permit the required compilations.
2. Regional differences for selected industries and perhaps major industrial groups for:

   (a) wage jobs per plant

   (b) wage jobs per unit output

(c) wage jobs per dollar output

(d) power per unit output
(Census of Manufactures, 1947)

3. Export sales data, by states and census regions, for selected industries and major industry groups.
(Census of Manufactures, 1947, and Census of Business, 1948)

Walter Isard
Merton J. Peck

Appendix B

Request for Support of a Project in the Field of Regional Economic Studies

Introductory

For many years economists have engaged in various types of studies of regional economic problems, either independently or in connection with organized programs under, for instance, bureaus of business and economic research at universities or under the aegis of federal or state or other developmental agencies. Members of other social sciences have been similarly engaged. Consequently, there is a considerable body of literature dealing with regional problems as well as a sizeable number of people engaged actively in research projects and programs. To date there has been no organized means for exchanging views among such people. A need has been felt and has been expressed a number of times for a systematic means of meeting and exchanging experiences with the hope and firm belief that the quality of the work in the area would be greatly improved as a result of such interchange. A year ago last Christmas a group of economists met informally in connection with the meetings of the American Economic Association. Last Christmas a similar informal meetings was held in Chicago, again in connection with the meetings of the American Economic Association. Attached herewith is a list of the people who attended the second meeting. Following this list appear the names of others who expressed an interest but were unable to attend the informal gathering.

During the informal meeting in Chicago at Christmas a small working committee was set up to explore the possibilities of obtaining funds to bring the committee together for systematic discussions of the problem in this field and to convene a larger group for more extended discussions. The members of this small *ad hoc* committee were:

- Morris E. Garnsey, University of Colorado, Boulder, Colorado
- E. T. Grether, University of California, Berkeley, California
- Werner Hochwald, Washington University, St. Louis, Missouri
- Glenn McLaughlin, National Security Resources Board, Washington, D.C.
- Philip Neff, University of California, Los Angeles, California
- Stefan H. Robock, Tennessee Valley Authority, Knoxville, Tennessee
- Harold Williamson, Northwestern University, Evanston, Illinois
- Walter Isard, Harvard University, Cambridge, Massachusetts (Chairman)

Another informal meeting was held in connection with the meeting of the Midwestern Economics Association April 20, 1951 in Milwaukee. A list of the participants of this meeting is appended.

A similar meeting is planned in connection with the two-day session of the Western Economics Association at Santa Clara University, Santa Clara, California in September.

At the annual Conference on Research in Income and Wealth at the National Bureau of Economic Research in New York City May 25 and 26 a session will be devoted to discussing a formal paper on "Regional and National Product Projections and Their Interrelations."

A large number of professional economists and agencies are currently engaged upon regional economic studies. In each section of the country projects are under way and occasionally there are evidences of some informal local or regional exchange of views and cooperation. For instance, at the April meeting of the American Association of Collegiate Schools of Business at Savannah, Georgia, some of the Southern deans met to discuss regional economic research in the South. On the Pacific Coast the Pacific Coast Board of Inter-Governmental Relations acts as a center for the states of California, Washington, and Oregon in a more or less informal manner. The governors of the three states have taken a strong interest in this organization together with the Federal Bureau of the Budget. Some economists in the intermountain states have met occasionally to discuss regional problems. Numerous federal agencies, of course, are engaged upon regional programs. The Council of Economic Advisers, under the leadership of Dr. Edwin G. Nourse, who was Chairman until recently, developed a strong interest in regional research. The banks of the Federal Reserve System typically have research staffs at work on regional problems. There are, of course, many active private agencies.

Many other evidences could be given of current interest in regional economic studies, or of past endeavors. No attempt is made here to make a complete report. In spite of the wide interest and activity, this type of analysis has not made its full contribution to the developments of the regions of this country and to the country as a whole. The present state of affairs and the immediate outlook from the standpoint of national security in our troubled world make it highly important that every effort be made to improve the planning and results of regional research activities.

Possible benefits from suggested program

It should be possible through group thinking and discussion to obtain the following benefits among others:

1. To improve the conceptual framework of regional economic studies.
2. To make suggestions to federal, state and other agencies for the provision of essential data where it is now lacking.

3. To indicate the most important problems and problem areas where research attention is needed. Too much effort is being devoted now to relatively unimportant and ephemeral problems instead of focusing upon fundamental issues. Some elements of superficiality stem from latent or active autarchic attitudes.

4. To develop suggestions for the improvement in methods of research and perhaps new approaches, through a critical examination of established methods and procedures as well as of new ones that are being proposed. For instance, one of the newer approaches which seems to have great potentialities is the regional input-output approach developed by Professor Wassily Leontief of Harvard University. An old approach which is receiving renewed attention is the analysis of geographical commodity flows. This type of approach was used effectively in the 1920's by the Department of Commerce. Unfortunately it was allowed to elapse except on a sporadic basis while resources were directed in less significant channels. If the data had been developed and the procedures sufficiently refined, the United States would have entered the period of World War II in much better condition so far as basic data and knowledge of the internal economy are concerned. There are strong evidences of interest now in this type of analysis, including the provision of better data. The Interstate Commerce Commission has made some sample waybill studies and also prepares tonnage data on the orientation and destination of freight movements by states and regions. In the meantime, some of the Federal Reserve Banks have become interested in the analysis of the monetary flows and transfers that parallel commodity movements. It is to be hoped that both types of analyses, broadened to include capital transfers, will eventually be correlated with a basic analysis of regional balances of payments. Some preliminary work along these lines has been done.

5. Possibly to advise concerning the development of soundly conceived and coordinated regional projects under various auspices.

6. An important by-product of organized cooperation in this field should be the development ultimately of better cooperation among the disciplines, especially in the social sciences. Geographers, political scientists, historians, members of law faculties, sociologists, psychologists, and others as well as economists are engaged to some extent on regional studies. There is a strong need ultimately for organized discussion among all such groups.

Specific proposals

It is proposed to begin with, a group of economists be convened for preliminary study and analysis. A part of the program of this group ultimately should be to make a recommendation for a broader inter-disciplinary approach.

The following requests are made as a means, in a preliminary way, of trying to bring the benefits indicated above to the field of regional economic analysis.

1. That a fund of $25,000.00 be provided for the initial work of the above committee plus others that it may invite to meet for preliminary discussions. These preliminary meetings should continue over a period of days, possibly as long as one week each. At these meetings the scope, problems, and needs of the field should be surveyed and a preliminary statement prepared. The fund of $25,000.00 would be used for travel expenses, subsistence, and secretarial or other assistance to expedite the work of the committee including the employment of a bibliographer to prepare a selected bibliography. An important aspect of the initial work should be the preparation of plans and an agenda for a more extended discussion by a larger group.
2. A fund of $25,000 for a larger representative gathering from all parts of the United States. This conference might take the form of a seminar running, say, two weeks. This larger group should include a few people outside of economics on a carefully selected basis.
3. A fund of $10,000 for the preparation of a report based upon the work of the committee and the seminar of the larger group and for the publication of a volume of essays dealing with the developments, problems, issues, needs, and approaches in the field of regional analysis. Undoubtedly, as a result of the organized discussions and the seminar, a volume of essays of high quality and importance could be prepared.

Final comment

It is assumed that one outcome of the discussions of the committee and of the larger group and of the preparation of the papers would be the development of suggestions for (1) the organized exchange of experience among economists as well as among members of other disciplines, and (2) useful research projects in terms of the national interest, and (3) refinements in analysis and approach which would greatly facilitate research.

The amounts proposed could be allocated in units subject to report rather than as a lump sum, if desired. For instance, the committee might be asked to make a preliminary report after its preliminary meetings before going ahead with the larger seminar. Matters would be expedited, however, if the total sum could be granted to begin with so that some momentum could be maintained throughout. The project as outlined is feasible and would undoubtedly move along as planned. At least two years would be required to follow the proposed course of action.

The initial committee need not be confined to the present working group which came into being somewhat spontaneously. It would, however, provide an established nucleus of interested persons with experience in regional research. If funds are granted, some agency such as the University of California or Harvard University, the National Bureau of Economic Research, or the Social Science Research Council would undoubtedly be willing to provide administration in accordance with established procedures. Personally, I should prefer the Social Science Research Council. I have taken the liberty of talking to Professor Harold E. Jones, the Pacific Coast representative in the Council. I am sure Professor Jones would be glad to advise in the matter.

E. T. Grether
University of California
May 24, 1951

Appendix C

Regionalism and American Economic History
Lee Benson (12/9/51)

During the past two decades American scholars have given increasing attention to the concept of regionalism. As a result, a considerable body of literature relative to that topic now exists in several disciplines. In American historical studies, however, the manifold references to regional or sectional phenomena have led to little *systematic* exploration of the possibilities that the concept of regionalism affords as a tool for historical analysis. At present it is hardly an overstatement to assert that almost as many varieties of regional interpretation exist as there are historians offering such interpretations. There is not even agreement concerning the actual spatial areas denoted by the usual terms of North, South, East, West, and combinations thereof. It does not, therefore, appear inappropriate to suggest that it is time for historians seriously to consider the regional concept, carefully to evaluate its utility for historical research, and co-operatively to embark on a program to maximize its usefulness.

In many ways American economic historians are peculiarly well qualified to initiate such a project. Economic theory has recently made marked advances in dealing with the economics of location and regional development. Unfortunately, these efforts are virtually unknown to historians of the non-economic breed, are usually presented in unfamiliar terms, and appear to have had scant impact upon empirical research. The converse is equally true: economic theorists do not appear to have made effective use of historical knowledge or methodology in the process of formulating and developing a set of general principles. Hence, with one foot in each camp, the economic historian would seem to be strategically situated. He is in a position to help bridge the gap between theorist and empiricist and thereby significantly contribute to the development of the regional concept.

Three considerations appear to strengthen the case for assigning regionalism a high priority among projects of interest to economic historians. In practice the latter find it difficult to avoid dealing with regional considerations. Secondly, other disciplines, such as economic theory, sociology, geography, ecology, etc., that have found regionalism to be a valuable tool in pursuing their particular fields of research, may reasonably look to economic historians to supply the data for studies of change over time. Finally, the concept of regionalism as a tool differentiates it from other areas of investigation because it offers the possibility of throwing *some* light upon *all* phases of American economic history. Essentially it is a methodological approach rather than a single research topic.

In view of these considerations the proposal is made that the committee on Research in Economic History establish an informal sub-committee—co-opting scholars of other disciplines as necessary—to evaluate the regional concept as an analytic tool for historical research, and to draw up a program to develop its utilization. If such a committee is formed—and stress is placed here on its informal, non-bureaucratic character—it is suggested that the representatives of various disciplines invited to participate in its work be men now actively engaged in regional studies.

The remainder of this memorandum presents in oversimplified and incomplete fashion possible lines of action which the sub-committee—if appointed—might consider.

At the outset it might be helpful to have an inventory taken of regionalism as it has been developed in the United States; i.e., the literature that it has produced, existing areas of agreement and disagreement, existing methods of handling regional analysis, etc. Emphasis should be placed upon the work of historians, but it will also be necessary to devote some attention to the development of the concept in other disciplines. However, because other social scientists have usually failed to concern themselves with *changes over time*, caution must be exercised in transferring the latters' schema to historical analysis.

If the experiences of other disciplines are any guide, the committee will be confronted by an initial methodological problem:

What is meant by the term "region"?

Can an area be considered a "region" for such varied purposes as economic, political, physiographic, and social differentiation? What criteria determine the size of a "region" for effective and meaningful handling of problems arising in separate contexts—economic, social, ecological, etc.? Similarly, like problems arise in determining the existence and boundaries of sub-regions and smaller units.

Secondly:

In so far as "regions" of a defensible character—size, differentiation on the basis of significant criteria, etc.—can be established relative to one point in time, should they be conceived as fixed through time? Do the criteria for regional delineation change over time? Do the number of regions—and smaller units—change?

A whole host of questions has arisen in connection with the fundamental methodological problem involved in delineating regional boundaries. Some scholars who stress the value of the regional concept as an analytic tool vigorously deny the possibility of marking off fixed geographic regions embracing a range of social phenomena. They argue that there must be as many regions as there are separate problems to be considered; one exponent of this view is on record as saying that perhaps "fixed regions" (areas embracing a range of social phenomena) like "class structures" are simply "folk words". In contrast to this atomistic approach, other scholars have designated a rigid set of regions which either implicitly or explicitly hold for the past, present, and future.

It is suggested here, however, that much of the discussion to date has been confusing, and, consequently, incapable of achieving persuasive conclusions. In part this stems from the static or narrowly deterministic thinking upon which it has been based, and in part from the atomistic concern with minute problems taken out of context from their total setting. To historically oriented researchers who shun narrow deterministic concepts, it would seem manifestly invalid to fix boundaries for American regions and expect them to hold true in 1800 A.D., 1900 A.D., 2000 A.D. Almost self-evidently, the number, area, characteristics, and *importance* of regions (including regional consciousness) change over time, particularly in as dynamic an economy as has existed in the United States. Similarly the criteria to be used in delineating regions can hardly be expected to be identical in the 18th, 19th, and 20th centuries. An atomistic approach which would construct an infinite number of regions to fit an infinitive number of problems is open to equally severe objections. It seems difficult to believe, for example, that anyone familiar with American history would deny that some definite relations exist between the natural environment of the South (however defined), its economic development, and its socio-cultural attitudes. What appears to be called for is a dynamic approach to regionalism, based upon the recognition that history is the process of continuity, unity, and change—and that generalizations derived from data pertaining to 1930 do not necessarily hold for 1830 or 1950.

Although the approach called for above is believed to be eminently sound, it is not necessary to accept it to begin a meaningful analysis of regional economic development in the United States. Logically, the first concrete step in the analysis might be the collection and classification of the *pertinent basic data* and their presentation in a convenient, systematic, and meaningful fashion. As yet that has not been done for the economic (political, social, cultural, etc.) history of the United States. By pertinent data is meant, for example, the geographic location of economic activity over time, i.e., where and how manufacturing, agriculture, commerce, finance, etc. were carried on at different periods. (By "how" is meant whether the activity occurs in factories or households, small farms or large farms, etc.). Obviously, the more inclusive, detailed, and accurate the data, the more definite the conclusions that can be drawn regarding the existence or non-existence of regions—and their precise contours. Although the questions raised above would still pose exceedingly nice problems of judgment, their resolutions could then proceed from the facts, not predilections or suppositions. In other words, what economic historians need to know *to begin* the discussion of regions and regionalism is the actual structure of the American economy as it has developed over time. Fortunately, the decennial Census since 1790 and various governmental and nongovernmental agencies have collected an enormous amount of the necessary data. But, just as agencies have collected an enormous amount of the necessary data, just as unfortunately the data are so thoroughly scattered and so unsystematically arranged that it is practically impossible to utilize them in spatial terms. Moreover, many valuable data are unpublished and cannot be sorted out by even the most diligent scholar working on his own. Hence it is suggested that the first major task for economic historians concerned with regionalism is the adoption of a program aimed at the presentation of the spatial structure of the American economy by Census decades since 1790, and (if feasible) at regular intervals before that date—as far as possible on a county basis.

No intention exists here of minimizing the tremendous difficulties inherent in such a program, nor is there any disposition to urge the program as a magic lantern to light the way to universal truths. On the contrary it is viewed merely as a first, but an indispensable first step if real progress is to be made relative to regionalism. Indeed, it might not be too much to assert that such a program has to be undertaken at some time if economic history is ever to realize its full potentialities. The details of the program need not be touched upon in this memorandum; essentially they are administrative, are not likely to produce wide differences in opinion, and are probably best handled as measures are taken to implement the project. It might also be stressed that this memorandum is not adopting a perfectionist approach, nor is insistence being voiced that all the pertinent data be assembled simultaneously, nor that all else must wait upon their publication. However, emphasis is placed upon the idea that publication of the data is basic if major advances are to be made in the study of American economic history; that it would be invaluable as a check upon the validity of existing hypotheses; and that it would be equally useful as a source of future hypotheses. In short, it seems reasonable to suppose that the more we know about what actually happened in American economic history—including where and when it happened—the better will we be able to explain why it happened, how it happened, and who caused it to happen.

The scale of the project, and the practical difficulties inherent in its operation, rule out the possibility of assembling all the pertinent data rapidly—if not for a considerable time to come. Hence, even more than usual, the aid of economic theorists and economic theory must be sought in the establishment of a system of priorities for the processing and publication of data. Just as the theorists appear to have been handicapped by an indifference to empirical research, empiricists have frequently floundered about in a morass of unrelated details. Cooperation between the two is eminently desirable if such a project is to be undertaken on any considerable scale of effort.

If the structure of the American economy is to be presented in the round, the effective utilization of data already collected in one form or another needs to be supplemented by other information. For the most part this information will have to be gathered by patient research of the kind painfully familiar to historians. For example, wage and profit differentials over time in identical or similar enterprises by geographic area; interest rates and availability of capital by area; ownership and control of enterprise by area; or by the existence and strength of farm, industrial, labor, commercial, financial organization by area (i.e., granges, chambers of commerce, trade unions, etc.). What is being called for here is not necessarily the causes of such differentials, but factual establishment of the differentials—if there were any.

Apart from knowledge of the actual spatial structure of the American economy by decades, it would be most desirable if economic historians knew the spatial distribution of natural resources (broadly considered) in terms of the existing technology (broadly considered). Distribution of natural resources is conceived of here on two levels: one, as we know the distribution today; two, as it was known at the time. Similarly, a distinction must be made in regard to technology: between knowledge by advanced personnel or firms in a given field, and implementation between knowledge by advanced personnel or firms in a given field, and implementation of the techniques or equipment on a mass scale. If such information were available, historians would then be able to contrast what might be called the potential structure of the economy with the actual structure; and attention could then be directed to explaining any marked disparities between the two structures. And the ranking of areas on the basis of realized potential might differ considerably from that based on absolute economic development. For example, the appraisal of achievement of Northern and Southern regions (however defined) might need to be modified. To a certain extent it may well be that such a disparate economic pattern was "natural," given considerations such as the availability of coal as the prime power source, topographic barriers to rail transportation in the South, strict limitations imposed upon Southern agriculture by climate and soil before the advent of scientific progress, etc.

Until now the memorandum has essentially been focused upon regionalism relative to what did happen, and what could have happened, given certain conditions. In a sense, although filled with enormous technical difficulties, these phases of inquiry are relatively the easiest in terms of satisfactory resolution. The major difficulties they present might really be viewed as mechanical and administrative; they call for efficient organization and adequate forces rather than for highly skilled, intelligent, and imaginative research. When attempts are made to answer questions involving the why, how, and who of regional economic development, a considerably more complex field is entered.

If there have been significant differences in regional economic development, why did these differences come about, how was it accomplished and who (i.e., what groups)—if anyone—was responsible for their appearance and their persistence? Did these regional differences stem from disparity in natural resources, divergent patterns of settlement, variations in entrepreneurial ability, disparities in sociocultural institutions and attitudes, etc., or from any combination of interactions among these and similar factors? Why did wage, interest, transportation, and profit rates differ? Why did migration really take place from one region to another? How can one explain the existence of different industrial-agricultural ratios, different degrees of mechanization? What has been the impact of these differences—in so far as they existed—on the development of trade unions, farmers' organizations, commercial organizations? Has the existence of regions contributed to or retarded the development of such institutions? Has there been significant export (or import) of capital on an interregional basis? If so, why, when, and how did it come about? Who were the people involved in these transactions? What institutions did they employ or create? Did capital exports—or other economic instruments of control—result in systematic exploitation of the people of one region by various groups and individuals in another region? Can it be said that there have been semi-imperialistic and semi-colonial regions in the United States?

The answers to these questions—and the numerous others that easily come to mind—obviously involve the entire range of social organization and social behavior. Although they are presented here primarily as problems in economic history, they are simultaneously problems in political, social, and cultural history. Progress toward their resolution will depend upon intelligent, imaginative, careful and lengthy research by scholars willing and able to overcome the barriers created by excessive specialization.

Obviously a long time will elapse before sufficient personnel and resources will be available to tackle all the questions noted above. Therefore, the proposal is made that three specific fields of inquiry be assigned high priority and that a detailed program be drawn up to encourage research in them. All three are extremely important, extremely complex, and all three are fundamental to the entire pattern of regional economic development, past, present, and future.

The first inquiry is a series of studies of the evaluation of the American railroad freight structure by spatial area or region (however determined) over time, and an analysis of the differentials—if any. It would include the level of *strategic* class and commodity rates, the classification of rates, the development of regional rate and classification territories, and of organizations to administer the territories. Rarely, if ever, are the historical aspects of these topics treated intensively and critically in the secondary works now available. The present author's studies in post–Civil War trunk line history lead him to conclude that too much emphasis cannot be placed upon the fact that research in this field has to be exceedingly thorough if it is to attain meaningful results. For example, if a study is made of the Southern rail rate structure, a really intensive inquiry would need to consider the question of actual ownership and control of Southern railroads. Were the roads the property of absentee-owners? Did these owners possess other property in the area served by their roads? Was their purpose in building up the territory to maximize profits on a long-term basis, or were they interested in short-term profits? Does any basis exist for the familiar charge that Southern railroad rates were deliberately designed to retard the industrial development of the region? If attempts were made to put such policies into effect, what groups supported and what groups opposed them? Granted the existence of such attempts, how successful were their proponents in attaining the desired ends? Were the roads operated to maximize profits for "inside managerial rings," or for the stockholders? To what extent were the stocks of the roads watered? Was capital difficult and expensive to procure? Were excessive profits made in the building of the roads? Did managers formulate constructive rate policies, were they well-trained for their jobs, were they efficient? In similar fashion, the other major factors involved (completed railroad mileage, topography, natural resources, traffic flows, pools and traffic associations, over-all economic development of area, water competition, governmental action, etc.) must be analyzed intensively and critically; informed acute skepticism might be the proper spirit in which to undertake such inquiries. The intricacies and pitfalls of the problem suggest that the most fruitful results would be gained by studies covering a relatively modest chronological and spatial range. But, along with this detailed analysis, an overall perspective would have to be maintained; the railroad rate structure developing throughout the country at large must form the background for such an analysis. Working together, transportation economists and railroad historians should be able to posit specific research projects which would throw considerable light upon the impact of rate structures upon regional development,

The second field of endeavor is even more complicated and extensive than the first: studies of interest rates and availability of capital by spatial area or region over time. Did significant regional differentials exist? To what extent was lack of capital or the existence of high interest rates a cause of retarded regional development; to what extent was it an effect? How did state and federal legislation influence the regional patterns? Were deliberate efforts made to keep interest rates high in different regions? Was control of capital used to manipulate the country's economic pattern in favor of one region as against other regions? Was federal legislation deliberately enacted to favor one region as against other regions? What were the channels through which capital was secured in the different regions? Did foreign investors plan an important role in the establishment of regional differentials? How did the regional pattern of settlement affect the regional interest rates and supply of capital? Of course, these questions need to be broken down still further to apply to specific spheres of economic activity, i.e., agriculture, manufacturing, commerce, transportation, etc. Again, it is suggested that specialists in this field—theorists and historians—work out a series of research projects cast in regional terms. Probably no other topic figures so prominently in American economic history; and probably about no other major topic does less reliable information exist. Until the gap is filled, much of the discussion of regional development must remain tentative and inevitably subject to unending revision. The "money question" runs all through American history; it appears to afford an excellent opportunity to test the utility of the regional concept as an analytic tool.

Finally, a series of studies is suggested which would examine the relationship between the locational patterns of specific industries over time, the spatial distribution of natural resources, and the impact of technological change. Such an approach might be effectively employed in analyzing the historical development of two important American industries; iron and steel, and textiles. Without neglecting other important considerations, the major focus of these studies would be upon the shifting locational pattern—or the failure to shift—in the light of natural resources and technology (both broadly considered). The emphasis upon resources and technology is merely a matter of degree; no disposition exists to disregard transportation, capital, labor, markets, entrepreneurship, etc., and the overall course of American historical development. Obviously, both industries can be studied from all these points of view; the technology and resources approach must take into account *all* factors affecting the location pattern and assign them proper weight in drawing final conclusions. Hence the suggestion is again made that the most fruitful results would be attained by intensive studies covering a relatively modest chronological and spatial range—studies undertaken in a critical spirit and satisfying all the canons of historical methodology.

Among the questions such studies might consider are these:

Did wide disparities exist between the potential regional development of these industries and their actual regional development (as defined above)? To what extent did the distribution of resources determine the distribution of plants? Did changes in industrial technology (combined with or separate from transportation advances) increase or decrease the locational pull of regional resources? Did increased knowledge of the existence or properties of various natural resources significantly affect locational patterns? Were organized and sustained efforts made on a regional basis to improve technology or discover new resource sites? Were efforts made to maintain regional locational patterns through patent control of technological innovations? If such efforts were made, how successful were they? Did heavy capital requirements handicap the utilization of improved technological processes by capital-importing regions?

To sum up, this memorandum has presented two proposals for economic historians to consider relative to regionalism. One is the collection and publication of basic data to provide a starting point for the resolution of fundamental questions regarding the existence and delineation of regions. The other is the launching of specific research projects in economic history constructed along regional lines. An overall, interdisciplinary planning committee seems surely to be called for as an essential first step. Out of its deliberations and out of the concrete activities that it might initiate, economic historians as a whole would gain an appreciation of the regional concept as a valuable tool in historical research.

Appendix D

Conference of the Regional Science Association 4–7 September 1961, Institute of Social Studies, Molenstraat 27, The Hague, Netherlands

List of Participants

	Name	Address
1.	Mr. Kebede Akalewold	Municipality of Addis Abeba, Ethiopia
2.	Mr. Anselin	Volderstraat 9, Gent, Belgium
3.	Prof. Roland Artle	Univ. of California, Berkeley, USA
4.	Mr. H. Baeyens	Froissartstraat 118, Brussels, Belgium
5.	Mr. A. R. Bakhtiar	c/o Plan Organization, Economic Bureau, Teheran, Iran
6.	Mr. A. Bassals	Unidad Esperanza 332, Mexico City, Mexico
7.	Mr. G. J. van den Berg	Geographical Institute, Utrecht, Netherlands
8.	Mr. P. Bernard	15 Avenue Bretteville, Neuilly s/Seine, France
9.	Mr. K. Bherwany	P.O. Shantinagar, Karachi 12, Pakistan
10.	Mr. M. Blazek	Ekonomicky Ustav, Prague, Czechoslovakia
11.	Mr. Th. Blumenthal	Mutschellenstrasse 120, Zurich, Switzerland
12.	Dr. W. Boes	Vilderstraat 20, Hasselt, Belgium
13.	Dr. F. Boesler	Drachenfelsweg 14, Beuel/Bonn, Germany
14.	Mr. K. A. Boesler	Hohenzollernstr. 24, Berlin-Wannsee, Germany
15.	Mr. H. C. Bos	Van Weberlaan 4, Rotterdam, Netherlands
16.	Mr. J. Boudeville	29 Rue Jasmin, Paris 13, France
17.	Mr. R. Bultynck	Voldersstraat 9, Gent, Belgium
18.	Mr. André Buyst	Voldersstraat 9, Gent, Belgium
19.	Miss V. Cao-Pinna	41 Via Chigi, Rome, Italy
20.	Prof. H. Chenery	Stanford University, California, USA

	Name	Address
21.	Dr. Walter Christaller	Jugendheim, Germany
22.	Mr. E. van Cleeff	Centraal Plan Bureau, Van Stolkweg 14, The Hague, Netherlands
23.	Mr. F. J. Croonenberghs	B.P. 368, Usumbura, Burundi, Ruanda-Urundi
24.	Mr. S. Czamanski	13/9 Henrietta Srold St. Haifa, Israel
25.	Prof. Dr. S. Dabčevic-Kucar	Novakova, Zagreb, Yugoslavia
26.	Mr. D. H. Davies	South Africa (Carpentierstraat 137c, The Hague, Netherlands)
27.	Mr. Y. Delcourt	71 Koning Albertlaan, Kesselo, Belgium
28.	Mr. J. Dooghe	Troonstraat 54, Brussels, Belgium
29.	Mr. Rudolph Doernach	Hochschule für Gestaltung, Ulm, Germany
30.	Mr. D. Dusseldorp	Herenstraat 25, Wageningen, Netherlands
31.	Mr. B. Einarsson	Ljosheimar 11, Reykjavik, Iceland
32.	Mr. Fillinger	Greyerstr. 32, Bern, Switzerland
33.	Mr. H. Font Viale Rigo	Urb. Sta. Cecilia Galles 12, Caracas, Venezuela
34.	Mr. J. Foërson	Holbertsgade 23, Copenhagen, Denmark
35.	Mr. I. Friščic	Brace Baruha 14, Belgrade, Yugoslavia
36.	Mr. V. Fuchs	477 Madison Ave., New York, USA
37.	Mr. S. Geronimakis	3 Amerikis Street, Athens, Greece
38.	Mr. Baolo Guidicini	Corticella 252, Bologna, Italy
39.	Mr. Niels Hansen	Volderstraat 9, Gent, Belgium
40.	Mr. L. Hattery	Am. University, Washington, USA
41.	Mr. A. Hauzer	Av. Louise 129 A, Brussels, Belgium
42.	Ir. W. C. A. van Heesewijk	Graafseweg 260, Den Bosch, Netherlands
43.	Miss C. Henriques	Banco de Fomento, Rua Braamcamp 5, Lissabon 1, Portugal
44.	Mr. C. Herbst	D. Burghelea 1, Bucharest, Romania
45.	Mr. M. Herbst	Hochschule für Gestaltung, Ulm, Germany
46.	Mr. W. Hochwald	Washington Univ., St. Louis 30, USA

	Name	Address
47.	Mr. Per Holm	Fleminggatan 62 B, Stockholm, Sweden
48.	Mr. S. Ilešic	Trstenjakova 9, Ljubljana, Yugoslavia
49.	Mr. S. Illeris	Peblinge dossering 16, Copenhagen K, Denmark
50.	Prof. W. Isard	Univ. of Pennsylvania, Philadelphia 4, USA
51.	Mr. B. P. Jørgenson	Revalsgade 10, Copenhagen V, Denmark
52.	Mr. B. Jovanovic	Kosmajska 39, Beograd, Yugoslavia
53.	Mr. T. Kano	5-cho, 6-jo dori, Sakai-shi, Osaka-fu, Japan
54.	Prof. Dr. R. Klöpper	Deutschherrenstrasse 40, Bad Godesberg, Germany
55.	Mr. G. de Koning	Rijksdiesnt v. h. Nationale Plan, The Hague, Netherlands
56.	Mr. J. Kruczala	Al. Stowackiego 11 B, m. 18, Krakow, Poland
57.	Mr. A Kühn	Hohenzollernstr. 11, Hannover, Germany
58.	Mr. H. Kuipers	Jasmijnstr. 76, The Hague, Netherlands
59.	Mr. A Kuklinski	Swietojanska 13, Warsaw, Poland
60.	Mr. J. W. Kurstjens	Ringbaan 475, Tilburg, Netherlands
61.	Prof. J. R. Lasuen	516 Muntaner, Barcelona, Spain
62.	Miss Dr. E. Lauschmann	Wirtschaftshochschule, Mannheim, Germany
63.	Mr. A. T. A. Learmonth	Univ. of Liverpool, England
64.	Prof. S. Leszczycki	Krakowskie Przedmiescie 30, Warsaw, Poland
65.	Mr. Jon Léons	Västmannagatan 49 III, Stockholm, Sweden
66.	Mr. P. K. Lindenbergh	Hoogeinde 2, Tiel, Netherlands
67.	Mr. S. Lombardini	Via Ansomio 26, Milan, Italy
68.	Mr. A. Lopez	Ned. Econ. Instituut, Rotterdam, Netherlands
69.	Mr. P. Maillet	11, Bd. Joseph II, Luxembourg
70.	Mr. P. N. Mathur	Gokhale Institute, Poona 4, India
71.	Dr. M. J. Meirleir	Rogiersplaats, Brussels, Belgium
72.	Mr. W. I. van Merhaeghe	9 Volderstraat, Gent, Belgium

	Name	Address
73.	Mr. J. B. Metha	Pakistan, 7, West Wharf Road, Karachi
74.	Mr. L. Mikkelsen	Kalkager 5, Hvidovre, Denmark
75.	Prof. G. Mirabella	Via Nigra 3, Palermo, Italy
76.	Mr. M. Miwa	Osaka City Univ., Ogimachi, Kitaku, Osaka, Japan
77.	Dr. N. L. Nicholson	Manotick, Ontario, Canada
78.	Mr. J. M. Nijhuis	Warmoeziershof 1, Hendrik-Ido-Ambacht, Netherlands
79.	Mr. R. H. Osborne	Crippshall, Univ. of Nottingham, England
80.	Mr. K. A. Ottesen	Byagerveg 19, Birkerød, Copenhagen, Denmark
81.	Mr. C. van Paassen	Geographical Institute, Utrecht, Netherlands
82.	Mr. C. Palacios	Circunvalacion 18, Blanes, Costa Brava, Spain
83.	Mr. K. R. Petshek	U.S.A.
84.	Mr. K. Pfromm	Hochschule für Gestaltung, Ulm, Germany
85.	Mr. J. C. Piperoglou	Athens Techn. Inst., 24 Strat. Syndesmou, Athens, Greece
86.	Mr. T. A. Reiner	Univ. of Pennsylvania, USA
87.	Mr. H. Rittel	Hochschule für Gestaltung, Ulm, Germany
88.	Mr. F. Rosenfeld	49, Rue du Fal d'Or, 's Gloud (seO), France
89.	Mr. Samaha	15 Sakakini Street, Cairo, U.A.R.
90.	Mr. J. Scheelbeek	Sarphatipark 41 IV, Amsterdam, Netherlands
91.	Dr. S. Schneider	Michaelshof, Bad Godesberg, Germany
92.	Mr. J. L. Servais	Rue de Louvain, Brussels, Belgium
93.	Mr. H. Shibli	c/o Planning Commission, Govt. of Pakistan, Karachi, Pakistan
94.	Mr. J. Sirotkovic	Laginjina 7, Zagreb, Yugoslavia
95.	Mr. G. H. Slotemaker de Bruino	Central Planning Bureau, The Hague, Netherlands
96.	Mr. B. P. Speetjens	Souavel. 4, Tilburg, Netherlands

	Name	Address
97.	Mr. J. A. Sporck	25 rue de l'Académie, Liège, Belgium
98.	Mr. M. H. v. d. Stichele	51 Doessich Steenweg, Kortrijk, Belgium
99.	Prof. G. Stolnitz	Indiana Univ., USA
100.	Prof. Jac. P. Thijsse	Institute for Social Studies, The Hague, Netherlands
101.	Mr. T. Tjalkens	Riouwstraat 178, The Hague, Netherlands
102.	Mr. R. Trias Fargas	Rambla de Cataluna 47, Barcelona-7, Spain
103.	Mr. W. Trzeciakowski	Langiewicza 2, Warsaw, Poland
104.	Mr. O. Tulippe	54 Quai Orban, Liège, Belgium
105.	Mr. F. Uemura	1-14, Kajiyamachi, Takamatsu, Japan
106.	Mr. E. L. Ullman	Univ. of Washington, Seattle, USA
107.	Mr. Norbert Vanhove	Voldersstraat 9, Gent, Belgium
108.	Mr. M. Verhulst	170, av. Paul Doumer, Rueil-Malmaison, France
109.	Dr. I. Vinski	Nemčiceva 2, Zagreb, Yugoslavia
110.	Mr. H. A. Visser	Dept. of Finance, Hollandia, N. Guinea
111.	Prof. A. C. de Vooys	Geographical Institute, Utrecht, Netherlands
112.	Mr. J. van Waterschoot	24, Luxemburgstr., Brussels, Belgium
113.	Prof. E. de Vries	Institute of Social Studies, The Hague, Netherlands
114.	Mr. V. Whitney	Univ. of Pennsylvania, Philadelphia, USA
115.	Mr. M. Wilmington	3441 85th Street, Jackson Heights, New York, USA
116.	Mr. E. Wirén	Hornsgatan 128, Stockholm, Sweden
117.	Mr. H. O. E. Whol	Pr. Margrietlaan 32, Voorschoten, Netherlands
118.	Dr. A. Wrobel	Hoza 5, Warsaw, Poland
119.	Mr. K. Wullich	Hasenbergsteige 73, Stuttgart-S, Germany
120.	Mr. W. Wurm	Hochschule für Gestaltung, Ulm, Germany
121.	Mr. H. Ziegler	Baltlam, Wankstrasse, Munich, Germany
122.	Mr. W. Hirsch	Washington Univ., St. Louis 30, Miss. USA

Appendix E

Annual Programs in the United States, 1958–1967 (EXCEPT 1963)

Program: Regional Science Association
December 27–29, 1958 Palmer House, Chicago, Illinois

SATURDAY, DECEMBER 27, 1958

9:30 A.M. Subject: *Gravity and Relative Income Potential Models*

Chairman: *Edward L. Ullman*, University of Washington

Papers:
1. Regional Employment and Population Projections via Relative Income Potential Models
   Walter Isard and David Bramhall, University of Pennsylvania
2. Gravity Models and Trip Predictive Theories in an Urban Region
   Morton Schneider, Chicago Area Transportation Study

Discussants: *Edward J. Taafe*, Northwestern University; *J. Douglas Carroll*, Chicago Area Transportation Study; *Gerald A.P. Carrothers*, University of Toronto

---

2:30 P.M. Subject: *Ecological Analysis of Regions*

Chairman: *Philip M. Hauser*, University of Chicago

Papers:
1. Service Industries and the Urban Hierarchy
   Otis Dudley Duncan, University of Chicago
2. Population Distribution and Economic Activity
   Beverly Duncan, University of Chicago
3. Variations in Industrial Composition with City Size
   Hal H. Winsborough, University of Chicago

Discussants: *Brian J. L. Berry*, University of Chicago; *John Cumberland*, University of Maryland

---

8:00 P. M. Subject: *Selected Regional Topics I*

Chairman: *Chauncy Harris*, University of Chicago

Papers: 1. Whither Regional Science?
Lloyd Rodwin, Massachusetts Institute of Technology

2. The Southern Appalachians as a Problem Social Area
Ruport Vance, University of North Carolina

Discussants: *Robert A. Kavesh*, New York University; Members of the Regional Science Seminar, University of Pennsylvania

---

SUNDAY, DECEMBER 28, 1958

9:30 A.M. Subject: *Use of Regional Science Techniques in Business and Industry*

Chairman: *Stefan H. Robock*, Committee of Economic Development

Papers: 1. Forecasting Regional Economic Activity
William H. Miernyk, Northeastern University

2. Regional Analysis As a Business Tool
Randall T. Klemme, Northern Natural Gas Company

3. Business Applications of Area Input-Output Analysis
Werner Z. Hirsch, Washington University

Discussants: *Abraham Gerber*, American Electric Power Service Corporation; *Paul H. Gerhardt*, Midwest Research Institute

---

2:30 P.M. Joint Session with the American Economic Association
Subject: *Regional Economics and Industrial Location*

Chairman: *Seymour E. Harris*, Harvard University

Papers: 1. Industrial Complex Analysis, Agglomeration Economies, and Regional Development
Eugene Schooler and Walter Isard, University of Pennsylvania

2. Changes in the Location of United States Manufacturing Since 1929
Victor Fuchs, Columbia University

Discussants: *Wilbur R. Thompson*, Wayne State University; *Paul G. Craig*, Ohio State University; *Richard Netzer*, Federal Reserve Bank of Chicago

---

5:00 P.M. *Business Meeting*, Regional Science Association

8:00 P.M. Subject: *Regional Planning and Development*

Chairman: *Lyle Craine*, University of Michigan

Papers: 1. Regional Planning: A Problem in Spatial Interaction
John Friedmann

2. Urbanization and Regional Development
Britton Harris, University of Pennsylvania

Discussants: *Charles L. Leven*, Iowa State College; *Joseph Airov*, Emory University; *Ronald Wonnacott*, University of Western Ontario

---

MONDAY, DECEMBER 29, 1958

9:30 A.M. Joint Session with the Econometric Society; Subject: *Interregional Linear Programming*

Chairman: *Robert Dorfman*, Harvard University

Papers: 1. A Spatial and Dynamic Growth Model
Edward Berman, Corporation for Economic and Industrial Research

2. Application to Location and Flow Determination in the Tomato Processing Industry
Robert Koch and Milton M. Snodgrass, Purdue University

Discussants: *Frederick T. Moore*, RAND Corporation; *Earle W. Orr, Jr.*, Northwestern University; *Martin Beckman*, Yale University

---

2:30 P.M. Subject: *Selected Regional Topics II*

Chairman: *Edward B. Espenshade*, Northwestern University

Papers: 1. Transport Inputs at Urban Residential Sites
Duane F. Marble, University of Washington

2. A Linear Programming Model of Urban Traffic Patterns
Howard W. Bevis, Chicago Area Transportation Study.

3. The Trade Balance of the Pacific Northwest
Richard Pfister, Dartmouth College

4. Space and Economic Theory
Melvin L. Greenhut, Florida State University

Program: Regional Science Association
December 27–29, 1959, Sheraton-Park Hotel, Washington, D.C.

SUNDAY, DECEMBER 27, 1959 HOTEL SHOREHAM

8:00 P.M. Joint Session with the American Statistical Association.
Subject: *Statistical Techniques in Regional Analysis*

Chairman: *Joseph L. Fisher*, Resources for the Future, Inc.

Papers: 1. A Statistical and Analytical Technique for Regional Analysis
Edgar S. Dunn, University of Florida

2. Minimum Requirements Approach to Urban Economic Base
Edward L. Ullman and Michael Dacey, University of Washington

Discussants: Edgar M. Hoover, Pittsburgh Regional Planning Association; *Richard Muth*, University of Chicago; *David Bramhall*, University of Pennsylvania.

MONDAY, DECEMBER 28, 1959 HOTEL SHERATON-PARK

9:30 A.M. Subject: *Local, Regional and National Impact of Disarmament*

Chairman: *Victor Roterus*, Office of Area Development, Department of Commerce

Panel:

1. *Charles L. Leven*, University of Pennsylvania
2. *Alfred C. Neal*, Committee for Economic Development
3. *Solomon Barkin*, Textile Workers Union of America
4. *Emile Benoit*, Columbia University
5. *Walter Isard*, University of Pennsylvania
6. *Neil Jacoby*, University of California at Los Angeles

2:30 P.M. Subject: *Urban Development and the Social Scientist*

Chairman: *Lyle Fitch*, City of New York

A. The Case: A Policy Decision Without the Help of Research
William L. Rafsky, City of Philadelphia

B. Research Panel

1. Political Science Approach
Robert Dahl, Yale University
2. Planning Approach
William Wheaton, University of Pennsylvania
3. Economics Approach
Morton Schussheim, Committee for Economic Development
4. Sociology Approach
Scott Greer, Northwestern University

C. The Problem Reviewed: An Effective Bridge Between Decision-Maker and Researcher?
Senator Joseph Clark, Pennsylvania

8:00 P.M. Subject: *Selected Regional Topics*

Chairman: *Robert A. Kavesh*, New York University

Papers: 1. Regional-Demographic Interaction: Some Issues and Examples
William S. Peters, Arizona State University.

2. The Connectivity of the Interstate Highway System
William L. Garrison, University of Washington

3. A Regional Weight Input-Output Table
T. Y. Shen, Federal Reserve Bank of Boston

TUESDAY, DECEMBER 29, 1959 HOTEL SHERATON-PARK

9:30 A.M. Subject: *The New York Metropolitan Region Study—A Basis for Policy-Making?*

Chairman: *Werner Hirsch*, Washington University

Panel: 1. Viewed from the Approach of Government Organization
Norton Long, Northwestern University

2. As Seen From the Angle of Land Use Planning
Fritz Gutheim, Congressional Staff, Washington, D.C.

3. In Terms of Public Economic Decision-Making
Kirk R. Petshek, City of Philadelphia

4. In Terms of Private Decision-Making
Warren Lindquist, Downtown Lower Manhattan Association

5. Viewed from the Transportation Angle
Robert Mitchell, University of Pennsylvania

Response: for the New York Study, *Raymond Vernon*, Harvard Business School

---

2:30 P.M. Subject: *Analysis of Underdeveloped Regions*

Chairman: *Lloyd Rodwin*, Massachusetts Institute of Technology

Papers: 1. Regional Factors in Planned Economic Development
Stefan Robock, Committee on Economic Development

2. An Inductive Approach to the Regionalization of Underdevelopment
Brian J. L. Berry, University of Chicago

Discussants: *Paul Rosenstein-Rodan*, Massachusetts Institute of Technology; *Wilfred Malenbaum*, University of Pennsylvania; *Allan Rodgers*, Pennsylvania State Univesity.

5:00 P.M. *Business Meeting*, Regional Science Association

8:00 P.M. *Presidential Address*

Chairman: *Robert B. Mitchell*, University of Pennsylvania

Paper: 1. The Nature and Scope of Regional Science
Walter Isard, University of Pennsylvania

WEDNESDAY, DECEMBER 30, 1959 HOTEL SHERATON-PARK

9:30 A.M. Joint Session with the Econometric Society
Subject: *Econometric Models in Regional Analysis*

Chairman: To be announced

Papers: 1. Aggregation in Regional Econometric Models
Wilbur R. Thompson and John Mattila, Wayne State University

2. A General Equilibrium Model of Production
Trade and Location,
Leon N. Moses, Northwestern University

3. Demand Approximation in Interregional Linear Programming
Benjamin H. Stevens, University of Pennsylvania

Discussants: *Charles Tiebout*, University of California at Los Angeles, *James Henderson*, Harvard University

---

2:00 P.M. Subject: *Selected Topics in Metropolitan Analysis*

Chairman: *Gerald A.P. Carrothers*, University of Pennsylvania

Papers: 1. A Topological Model of Consumer Space Preferences
David L. Huff, University of Washington

2. A General Theory of the Urban Land Market
William Alonso, Harvard University

3:30 P.M. Joint Session with Econometric Society
Subject: *Migration of Labor*

Chairman: *Richard A. Easterlin*, University of Pennsylvania

Papers: 1. Factors Influencing the Net Migration of Labor from Agriculture
Charles H. Berry, Cowles Commission for Research in Economics at Yale University

2. An Investigation of the Relationship between Migration and Income in the United States
Larry Sjaastad, University of Chicago

Discussants: *Stanley Lebergott*, Bureau of the Budget; *Bernard Okun*, Princeton University

Seventh Annual Meetings
Chase-Park Plaza Hotel, St. Louis, Mo.
December 28–30, 1960

WEDNESDAY, DECEMBER 28, 1960

9:30 A.M. Subject: *Regional Accounts and Income Analysis*

Chairman: To be announced

Papers:
1. A Moneyflows Analysis of Metropolitan Saving and Investment
   Charles L. Leven, University of Pennsylvania
2. Industrialization and Income Inequality—Recent U.S. Experience
   Eugene Smolensky, Haverford College
3. Interregional Social Accounting in Relation to Other Regional Science Techniques
   David Bramhall, The Johns Hopkins University

Discussants: To be announced

2:30 P.M. Joint Session with the American Economics Association

Subject: *Economic Analysis of Urban Problems*

Chairman: *Sherman J. Maisel*, University of California

Papers:
1. Intra-Urban Location Theory
   Charles M. Tiebout, University of California at Los Angeles
2. Contrasts in Agglomeration—New York and Pittsburgh
   Benjamin Chinitz, University of Pittsburgh
3. Pricing Policies in Urban Redevelopment
   Louis Winnick, New York State Commission on Economic Expansion

Discussants: *Barbara R. Berman*, Harvard University; *Britton Harris*, University of Pennsylvania; *Irving Morrissett*, Purdue University

---

8:00 P.M. Subject: *Location Studies*

Chairman: To be announced

Papers:
1. Role of the Community as a Factor in Industrial Location
   Vernon W. Rutten and T. W. Wallace, Purdue University
2. Demand for Tourist's Goods and Services in a World Market: The Influence of Location and Other Factors
   Harold Guthrie, University of Kentucky
3. On Production and Urban Traffic

Discussants: To be announced

THURSDAY, DECEMBER 29, 1960

9:30 A.M. Joint Session with the Econometric Society

Subject: *Models of Urban Form and Structure*:

Chairman: *William L. Garrison*, Northwestern University

Papers:

1. An Economic Model of the Utilization of Urban Land for Residential Purposes
   Lowdon Wingo, Jr., Resources for the Future
2. The Spatial Structure of the Housing Market
   Richard F. Muth, University of Chicago
3. Urban Land Values and the Measurements of Benefits of Urban Highway Construction
   Herbert Mohring, Northwestern University

Discussants: To be announced

12:30 P.M. Council Meeting, Regional Science Association

2:30 P.M. Subject: *Behavioral Models in Regional Analysis*

Chairman: To be announced

Papers:

1. Ecological Characteristics of Human Behavior
   David L. Huff, University of California at Los Angeles
2. An Application of Game Theory to a Problem in Location Strategy
   Benjamin H. Stevens, University of Pennsylvania
3. Graph-Theoretic Techniques of Regionalization—An Empirical Example
   Michael F. Dacey, University of Pennsylvania, and *John D. Nystuen*, University of Michigan

Discussants: To be announced

5:00 P.M. Business Meeting, Regional Science Association

8:00 P.M. Presidential Address

Chairman: *Edward L. Ullman*, Washington University

Paper: A City Planner Looks at Regional Science
Robert B. Mitchell, University of Pennsylvania

FRIDAY, DECEMBER 30, 1960

9:30 A.M. Joint Session with the Econometric Society

Subject: *Regions and Resources*

Chairman: *William Vickery*, Columbia University

Papers:
1. Water and Welfare
   Robert Dorfman, Harvard University
2. Regional Fiscal Impact of Local Industrial Development
   Werner Hirsch, Washington University
3. Welfare Basis of Benefit-Cost Analysis
   John Krutilla, Resources for the Future

Discussants: To be announced

9:30 A.M. Subject: *Urban Policy: Does Economic Research Help?*

Chairman: *John H. Nixon*, Committee for Economic Development

Paper: Can Policy Decisions Be Made More Rational by Techniques of Economic Analysis?
Benjamin Chinitz and Melvin Bers, Pittsburgh Area Economic Study

Panel:
1. Economic Studies and Economic Policy
   Richard Netzer, Federal Reserve Bank of Chicago
2. Planning, Policies and the Social Sciences
   John W. Dyckman, University of Pennsylvania
3. Economic Techniques and Political Realities
   Robert Wood, Massachusetts Institute of Technology
4. Useful Tools for Consulting Governments
   Philip Hammer, Hammer Associates

2:30 P.M. Subject: *Modern Transportation and Metropolitan Policy: Projection, Prediction or Plan?*

Chairman: *Henry Fagin*, Penn-Jersey Transportation Study

Papers:
1. Detroit—The Future Seen by Transportation Plan and Regional Plan: One Vision or Two?
   Robert Hoover, Wayne State University
2. Chicago—Can Policy Change the Flow of Urban Growth?
   Harold Mayer, University of Chicago
3. Washington, D.C.—The Transportation Plan and Darwinian Evolution
   Harvey Perloff and Lowdon Wingo, Jr., Resources for the Future

Discussants: *Frederick Gutheim*, Washington Metropolitan Center; *Lyle Fitch*, City Administrator, New York City; *Paul Oppermann*, Northwestern Illinois Metropolitan Planning Commission

Eighth Annual Meetings
Biltmore Hotel, New York, N.Y.
December 27–29, 1961

WEDNESDAY, DECEMBER 27, 1961

9:30 A.M. Subject: *Behavioral Studies in Regional Science*

Chairman: To be announced

Papers: 1. On Projecting Behavior for Regional Analysis
Walter Isard and Michael F. Dacey, University of Pennsylvania

2. The Household Sector in Location Theory
David N. Milstein, University of Michigan

Discussants: To be announced

2:30 P.M. Subject: *The Property Tax and Urban Land Use*

Chairman: *Jesse Burkhead*, Syracuse University

Papers: 1. The Property Tax Base and the Pattern of Local Government Expenditures: The Influence of Industry
John J. Carroll and Seymor Sacks, New York State Department of Audit and Control

2. The Property Tax and Alternate Patterns of Urban Development
Dick Netzer, Regional Plan Association

Discussants: *William Vickrey*, Columbia University; *Richard U. Ratcliff*, University of Wisconsin; *Jerome Pickard*, Urban Land Institute; *Lynn A. Stiles*, Federal Reserve Bank of Chicago

---

8:00 P.M. Subject: *Studies of the Urban Region*

Chairman: To be announced

Papers: 1. Transportation and the Spatial Structure of an Urban Area
Leon Moses, Northwestern University

2. A Simulation Model of Urban Settlement and Migration
Richard L. Morrill, University of Washington

Discussant: *Robert Kavesh*, New York University

THURSDAY, DECEMBER 28, 1961

9:30 A.M. Subject: *The Location of Retail Activities*

Chairman: *William L. Garrison*, Northwestern University

Papers: 1. Retail Location and Consumer Behavior
Brian J. L. Berry, H. Gardiner Barnum and Robert Tennont, University of Chicago

2. Spatial Behavior of a Dispersed Non-Farm Population
Edwin Thomas, State University of Iowa

Discussants: *Benjamin H. Stevens*, University of Pennsylvania; *Dennis Durden*, Larry Smith and Company

---

2:00 P.M. Subject: *Contributed Papers*

Chairman: To be announced

Papers: 1. Problems in the Construction of Regional Models
Michael B. Teitz, University of Pennsylvania

2. Regional Growth Models for Rural Economic Development
Wilbur R. Maki, Iowa State University, and *Yien-I Tu*, University of Saskatchewan

---

3:30 P.M. Subject: *Presidential Address*

Chairman: *Edgar M. Hoover*, Pittsburgh Regional Planning Association

Paper: The Nature of Cities Reconsidered
Edward L. Ullman, University of Washington

5:00 P.M. Business Meeting, Regional Science Association

8:00 P.M. Subject: *The Location of Residential Areas*

Chairman: To be announced

Papers: 1. Social and Economic Factors Affecting the Location of Residential Neighborhoods
Theodore R. Anderson, State University of Iowa

2. The Journey-to-Work as a Determinant of Residential Location
John F. Kain, the RAND Corporation

Discussant: *John D. Nystuen*, University of Michigan

FRIDAY, DECEMBER 29, 1961

9:30 A.M. Subject: *Regional Analysis in Economic History*

Chairman: *Douglas C. North*, University of Washington

Papers: 1. The Influence of the Intraregional Railroad on the Development of American Agriculture During the Nineteenth Century
Robert Fogel, University of Rochester

2. Exports and Economic Growth: The Pacific Northwest Experience 1880–1960
James Tattersall, University of Oregon

Discussants: *William N. Parker*, University of North Carolina; *Henry Broude*, Yale University

---

2:30 P.M. Joint Session with the Econometric Society

Subject: *Regional Econometric Studies*

Chairman: *Duane F. Marble*, University of Pennsylvania

Papers: 1. Factors Influencing Inter-Regional Differences in the "Mix" of Telephone Central Office Equipment
John A. Carlson, Princeton University

2. A California Intersectoral Output Flows Study
W. Lee Hansen and Charles M. Tiebout, UCLA

3. Production Functions, Establishment Size and Labor Quality
Phillip Nelson, New School for Social Research

Discussants: *Richard B. Maffei*, M.I.T.; *Martin J. Gerra*, U.S.D.A.; *Murray Brown*, University of Pennsylvania

---

Ninth Annual Meetings
The Penn-Sheraton, Pittsburgh
December 27–29, 1962

THURSDAY, DECEMBER 27, 1962

9:30 A.M. Subject: *Regional Resource Management*

Chairman: *Marion E. Marts*, University of Washington

Papers: 1. Use of Effluent Charges in Achieving Optimal Regional Waste Disposal Systems
Allen Kneese, Resources for the Future, Inc.

2. Regional Differentiation in Flood Plain Management
Robert Kates, Clark University

Discussants: *C. H. J. Hull*, The Johns Hopkins University

9:30 A.M. Subject: *Contributed Papers I*

Chairman: *Morgan Thomas*, University of Washington

Papers: 1. Measures of Regional Interchange
William Peters, Arizona State University

2. Regional Economic Policy with Special Reference to Venezuela
John Friedman, Massachusetts Institute of Technology

3. A Regional Analysis of the Demand for Hired Agricultural Labor
G. Edward Schuh and John R. Leeds, Purdue University

2:30 P.M. Subject: *Problems of Spatial Organization*

Chairman: *Edward L. Ullman*, University of Washington

Papers: 1. Some Organizational Concepts for a System of Regions
Walter Isard, University of Pennsylvania

2. Organizing Regional Investment Criteria
Thomas A. Reiner, University of Pennsylvania

Discussants: To be announced

8:00 P.M. Panel: *The Training of Regional Scientists*

Chairman: *Scott Keyes*, University of Illinois

Panel Members: *William L. Garrison*, Northwestern University; *David L. Huff*, University of California, Los Angeles; *Benjamin H. Stevens*, University of Pennsylvania; *Melvin M. Weber*, Resources for the Future, Inc.

FRIDAY, DECEMBER 28, 1962

9:30 A.M. Subject: *Regional Econometric Studies* (joint with Econometric Society)

Chairman: *Benjamin Chinitz*, University of Pittsburgh

Papers: 1. The Redistribution of Employment and Population within the Largest Metropolitan Areas
John F. Kain and John F. Niedercorn, The RAND Corporation

2. A Computer Model of the Spatial Structure of the Pittsburgh Economy
Ira S. Lowry, Carnegie Institute of Technology

3. Three Econometric Approaches to Population Adjustment
George S. Tolley and Gordon Sanford, North Carolina State College

Discussants: *Barbara Berman*, Brandeis University; *Brian J. L. Berry*, University of Chicago; *V. V. B. Rao*, Hyderabad, India

---

2:00 P.M. Panel: *Local, Regional, and National Aspects of Disarmament—The Data and Techniques Available for Research*

Chairman: To be announced

Panel Members: *Walter Isard*, University of Pennsylvania; *James Ganschow*, University of Texas; *Edgar Dunn*, U.S. Department of Commerce; *Charles Leven*, University of Pittsburgh; *Charles Tiebout*, University of Washington; *Robert Steadman*, Department of Defense

---

4:00 P.M. Presidential Address

Chairman: *William L. Garrison*, Northwestern University

Paper: Whence Regional Scientists?
Edgar M. Hoover, University of Pittsburgh

5:00 P.M. Business Meeting, Regional Science Association

8:00 P.M. Subject: *Spatial Diffusion Studies*

Chairman: *Edwin Thomas*, Arizona State University

Papers:

1. An Approach to the Direct Measurement of Community Mean Information Fields
Duane F. Marble, University of Pennsylvania
2. The Distribution of Migration Distances
Richard L. Morrill, University of Washington
3. Problems in Computer Simulation of Diffusion
Forrest R. Pitts, University of Oregon

Discussant: *Robert Mayfield*, Texas Christian University

SATURDAY, DECEMBER 29, 1962

9:30 A.M. Subject: *Problems of the Urban Region*

Chairman: *Herman G. Berkman*, New York University

Papers:

1. A Quantitative Study of Urban Reconstruction: The Rotterdam Case
Leland S. Burns, University of California, Los Angeles
2. Simulation and Urban Renewal
Benjamin Chinitz, University of Pittsburgh
Wilbur A. Steger, The Consad Corporation
3. Forecasting Public Education Expenditures in Local Areas
Richard Kosobud, University of Michigan

Discussants: *William N. Kinnard, Jr.*, University of Connecticut; *David Milstein*, University of Michigan

---

2:00 P.M. Subject: *Contributed Papers II*

Chairman: *Robert Nunley*, University of Kansas

Papers:
1. Spatial Variables and CBD Sales
   Ronald Boyce, University of Illinois
2. Prediction of the Incidence of Urban Residential Blight
   Fred E. Case, University of California, Los Angeles
3. On the Measurement of Economic Development Using Scalogram Analysis
   Magdi M. El-Kammash, North Carolina State College

Tenth Annual Meetings
The University of Chicago Center for Continuing Education
November 15–17, 1963

Presented in the Main Text, pp. 154–156

Eleventh Annual Meetings
University of Michigan, November 13–16, 1964

FRIDAY, NOVEMBER 13, 1964

2:00 P.M.–8:00 P.M. Registration—Third Floor Corridor

3:00 P.M. Classical Music Session—Third Floor Conference Room

Organizer: *Edgar M. Hoover*, University of Pittsburgh

8:00 P.M. Subject: *Ph.D. Theses*

Chairman: *Richard L. Meier*, University of Michigan

Papers:
1. Decision Procedures Affecting Land Use and Migration
   Richard Duke, Michigan State University
2. Quantitative Subjective Analysis in Urban Ecological Systems
   George L. Peterson, Northwestern University

Discussion from the Floor

SATURDAY, NOVEMBER 14, 1964

8:30 A.M. Registration—Third Floor Corridor

9:30 A.M. Subject: *Regional Science Theory and Models*

Chairman: (to be announced)

Papers: 1. Existence of an Equilibrium for a Social System
Walter Isard, University of Pennsylvania

2. Behavioral Aspects of the Decision to Migrate
Julian Wolpert, University of Pennsylvania

Discussants: *Roland Artle*, University of California, Berkeley; *Forrest R. Pitts*, University of Pittsburgh

---

2:30 P.M. Subject: *Regional Economic and Development Analysis*

Chairman: *Charles L. Leven*, University of Pittsburgh

Papers: 1. The Functional Economic Area: Delineation and Implications for Economic Analysis and Policy
Karl Fox, Iowa State University

2. Interregional (Intra-State) Aspects of a State Development Plan
John W. Dyckman, University of California, Berkeley

Discussants: *Lyle Craine*, University of Michigan; *Morgan Thomas*, University of Washington

---

5:00 P.M.–6:00 P.M. Council Meeting

8:00 P.M. Subject: *Statistical Procedures and Quantitative Methods*

Chairman: *Brian J. L. Berry*, University of Chicago

Papers: 1. Statistical Regularities in Economics and Regional Science: Some New Methods of Testing
Richard E. Quandt, Princeton University

2. Integration of Statistical and Cartographical Analysis
Kazimierz Dziewonski, Institute of Geography, Polish Academy of Sciences

3. Computation of the Correspondence of Geographical Patterns
Waldo R. Tobler, University Michigan

Discussion from the Floor

SUNDAY, NOVEMBER 15, 1964

8:20 A.M.–9:20 A.M. Early-Bird Session: *Ph.D. Theses*

Chairman: *John Nystuen*, University of Michigan

Papers: 1. Optimal Spatial Patterns of Production and Decision-Making
Tze H. Tung, Colorado State University

2. Predicting the Viability of Farm Trade Centers in the Great Plains
Gerald Hodges, University of Toronto

Discussion from the Floor

9:30 A.M. Subject: *Residential Patterns and Land Use Analysis*

Chairman: John Friedmann, Massachusetts Institute of Technology

Papers: 1. Simulation of Urban Residential Patterns
Richard Morrill, University of Washington

2. Intensity of Residential Land Use with Particular Reference to Chicago
Richard F. Muth, University of Chicago

Discussants: *Barbara Berman*, Brookings Institution, Washington, D.C.; *Cicely W. Blanco*, Kent State University

12:30 P.M. Lunch; Subject: *The Ph.D. Program and Other Developments in Regional Science*

Chairman: *Benjamin H. Stevens*, University of Pennsylvania

2:30 P.M. Subject: *Transportation Models and Analysis*

Chairman: *Duane Marble*, Northwestern University

Papers: 1. Optimal Interregional Allocation of Transport Investment
Gary Fromm, Brookings Institution, Washington, D.C.

2. Experiments in Simulating Urban Travel
Aaron Fleisher, Massachusetts Institute of Technology

Discussants: *Harold J. Barnett*, Washington University, St. Louis; *Ronald Miller*, University of Pennsylvania

5:00 P.M.–6:00 P.M. Business Meeting

7:00 P.M. Dinner and Presidential Address

Chairman: *William L. Garrison*, Northwestern University

Address: *William L. C. Wheaton*, University of California, Berkeley

9:30 P.M. Jazz Session

Organizer: *Benjamin H. Stevens*, University of Pennsylvania

MONDAY, NOVEMBER 16, 1964

8:30 A.M.–9:20 A.M. Early-Bird Session: *Ph.D. Theses*

Chairman: (to be announced)

Papers:
1. Estimation of the Demand for Transportation
   Eugene D. Perle, Indiana University
2. Spatial Interaction Models and Processes
   Herman Porter, Northwestern University

9:30 A.M. Subject: *Program of the New Division of Regional Economics, Office of Business Economics, U.S. Department of Commerce*

Chairman: *Edgar S. Dunn, Jr.*, Resources for the Future, Inc.

Papers:
1. Regional Change in a National Setting—A Formal System of Analysis
   Lowell D. Ashby, Division of Regional Economics
2. Regional Change in a National Setting—The Base Bias Problem and Other Formal Characteristics
   John M. Mattila, Wayne State University
3. Summary of Results of the Current Program
   Edgar S. Dunn, Jr., Resources for the Future, Inc.
4. Future Plans of the Division of Regional Economics
   Robert E. Graham, Division of Regional Economics

Twelfth Annual Meetings
Annenberg School of Communications, University of Pennsylvania
November 12–14, 1965

FRIDAY, NOVEMBER 12, 1965

8:30 A.M. Registration

10:00 A.M. Subject: *New Concepts and Theory in Regional Science*

Chairman: *Robert B. Mitchell*, University of Pennsylvania

Papers: 1. Game Theory Reconsidered: Resolution of Conflicts Among Regions of a System
Walter Isard and Tony E. Smith, University of Pennsylvania

2. The Topology of a Socio-Economic Terrain and Spatial Flows
William N. Warntz, American Geographic Society

Discussants: *William A. Alonso*, Harvard University; *David L. Huff*, University of Kansas

---

2:30 P.M. Subject: *Urban Transportation and Land Use*

Chairman: *J. Douglas Carroll, Jr.*, Tri-State Transportation Committee, New York City

Papers: 1. Interregional Migration and Economic Development
Leon N. Moses and Ralph E. Beals, Northwestern University, and *Mildred Levy*, University of Illinois, Chicago

2. Land Use Data Collection Systems: The Problem of Unification
Michael B. Teitz, University of California, Berkeley

Discussants: *John F. Kain*, Harvard University; *Eugene Smolensky*, University of Chicago

---

5:00 P.M.–6:00 P.M. Coffee Hour

5:00 P.M.–6:00 P.M. Council Meeting

7:30 P.M. Subject: *Spatial Patterns and Flows*

Chairman: *Britton Harris*, University of Pennsylvania

Papers: 1. Spatial Demand Theory and Monopoly Price Policy
Benjamin H. Stevens and C. Peter Rydell, University of Pennsylvania

2. Boundary Shapes and Transfer Problems
John D. Nystuen, University of Michigan

3. A Markovian Policy Model of Interregional Migration
Andrei Rogers, University of California, Berkeley

Discussants: *Tze H. Tung*, Colorado State University; *Julian Wolpert*, Michigan State University

---

SATURDAY, NOVEMBER 13, 1965

8:20 A.M.–9:20 A.M. Early-Bird Session: *Ph.D. Theses*

Chairman: *Peter Gould*, Pennsylvania State University

Papers: 1. The Employment Structure of Indian Cities, 1951–61
George Stoner, Jr., Bradley University

2. Interregional Welfare Equalization and Economic Efficiency
Koichi Mera, Harvard University

3. Locational Decision Factors in a Producer Model of Residential Development
Edward J. Kaiser, University of North Carolina

9:30 A.M. Panel: *Area and Regional Development Problems and Policy*

Chairman: *Gordon E. Reckord*, Area Redevelopment Administration, Washington, D.C.

Panelists: *John R. Meyer*, Harvard University; *Leland S. Burns*, University of California, Los Angeles; *Norton S. Ginsburg*, University of Chicago; *David Houston*, Pennsylvania State University

---

12:30 P.M. Lunch Session, Subject: *The Ph.D. Program and Graduate Study in Regional Science*

Chairman: *Benjamin H. Stevens*, University of Pennsylvania

2:30 P.M. Subject: *Regional Interindustry Models and Analysis*

Chairman: *Robert W. Eisenmenger*, Federal Reserve Bank of Boston

Papers: 1. A Regional Interindustry Model for Analysis of Development Objectives
John H. Cumberland, University of Maryland

2. Empirical Problems in the Implementation of the Maryland Interindustry Study
Robert G. Kokat, University of Maryland

3. Interregional Feedbacks Effect in I-O Analysis: Some Experimental Results
Ronald E. Miller, University of Pennsylvania

Discussants: *Ira S. Lowry*, Rand Corporation; *William H. Miernyk*, University of Colorado

---

5:00 P.M.–6:00 P.M. Business Meeting

5:00 P.M.–6:00 P.M. Student Coffee Hour

7:00 P.M. Dinner and Presidential Address

Chairman: *William L. C. Wheaton*, University of California, Berkeley

Address: Regional Allocation of Investment in the Public Sector: Economic Planning in a Capitalist Society
Charles L. Leven, Washington University, St. Louis

9:30 P.M. Jazz Session

Organizer: *Benjamin H. Stevens*, University of Pennsylvania

SUNDAY, NOVEMBER 14, 1965

8:20 A.M.–9:20 A.M. Early-Bird Session: *Ph.D. Theses*

Chairman: *Benjamin Chinitz*, University of Pittsburgh

Papers:
1. A Comparative Evaluation of Gravity and Systems Theory Models for Statewide Recreational Traffic Flows
Jack B. Ellis and Carlton S. Van Doren, University of Waterloo and Ohio State University
2. The Effect of Direction and Length of Person Trips on Urban Travel Patterns
David E. Boyce, Battelle Memorial Institute
3. Urban Renewal: Project Design and Analysis
David A. Page, U.S. Bureau of the Budget

9:30 A.M. Subject: *New Approaches to Regional Development*

Chairman: *Joseph L. Fisher*, Resources for the Future, Inc.

Papers:
1. On Activity Analysis Models in Regional Development Planning
Robert G. Spiegelman and Mordecai Kurz, Stanford Research Institute
2. Strategies for Regional Economic Development
*Stefan H. Rob*ock, Indiana University

Discussants: *Morgan Thomas*, University of Washington, Seattle; *Hans-Wilkin von Borries*, Institute of Economics and Social Science, Stuttgart-Hohenheim, West Germany

---

2:30 P.M. Subject: *Techniques in Conversion from Defense to Non-Defense Activities at the Regional and Local Level* (meeting organized by Peace Research Society (International), jointly sponsored by the Regional Science Association)

Chairman: *Harold J. Barnett*, Washington University, St. Louis

Papers: 1. The Philadelphia Interindustry Study: The Interrelations of Defense and Non-Defense Sectors
Eliahu Romanoff, Harvard University; *Thomas Langford, and other staff*, Philadelphia Impact Study, University of Pennsylvania

2. A Model for the Projection of Regional Industrial Structure, Land Use Patterns and Conversion Potentialities
Walter Isard and Stanislaw Czamanski, University of Pennsylvania

Discussants: Marshall Wood, National Planning Association, Washington, D.C.; *Wilbur R. Thompson*, Wayne State University

---

8 P.M. Subject: *Empirical Studies Relating to Conversion from Defense to Non-Defense Activities* (jointly sponsored with the Peace Research Society International)

Chairman: *William L. Garrison*, Northwestern University

Papers: 1. Shifting from Defense to Non-Defense Spending: The Implications for the Regional Distribution of Per Capita Income
Murray Weidenbaum, Washington University, St. Louis

2. Interregional Flows of Defense-Space Awards
Gerald J. Karaska, University of Pennsylvania

Discussants: *Hugh Knox*, University of North Carolina; Emile Benoit, Columbia University

---

Thirteenth Annual Meetings
Washington University (St. Louis) and Clayton, Missouri
November 4–6, 1966

FRIDAY, NOVEMBER 4, 1966

8:30 A.M. Registration

10:00 A.M. Subject: *Equilibrium Analysis*

Chairman: *Harold Barnett*, Washington University

Papers: 1. On a Synthesis of General Equilibrium and Game Theory for a Multi-Regional System
Walter Isard and Tony E. Smith, University of Pennsylvania

2. Solutions of Generalized Locational Equilibrium Models
Leon Cooper, Washington University

Discussant: *Michael B. Teitz*, University of California (Berkeley)

12:00 Noon–3:00 P.M. Lunch and other activities at Washington University

---

3:00 P.M. Subject: *Location, Rent and Transportation*

Chairman: *Edgar M. Hoover*, University of Pittsburgh

Papers:
1. A Reformulation of Classical Location Theory and Its Relation to the Theory of Rent
   William Alonso, Harvard University
2. A Model of an Integrated Transportation Network
   Allen Scott, University of Pennsylvania

Discussants: *Leon Moses*, Northwestern University; *Duane Marble*, Northwestern University

---

6:00 P.M. Meeting of the Regional Science Association Council

---

8:00 P.M. Subject: *Interregional Models and Analysis*

Chairman: *Gerald A. P. Carrothers*, Central Mortgage and Housing Corporation (Canada)

Papers:
1. Linear Programming Model for Agriculture
   Richard Howes
2. Interregional Fiscal-Policy Theory and Regional Full Employment
   Joseph Airov, Emory University

Discussion from the Floor.

SATURDAY, NOVEMBER 5, 1966

8:20 A.M.–9:20 A.M. Early-Bird Session: *Ph.D. Papers*

Chairman: *Gerald J. Karaska*, Syracuse University

Papers:
1. An Analysis of Industrial and Regional Distribution of NASA Subcontracts
   Robert Bohm, Washington University
2. Economic Base and Multiplier Approaches to Regional Income Generation
   Richard T. Pratt, University of Utah
3. Patterns of Industrial Change and Entrepreneurial Adjustments
   Gunter Krumme, University of Hawaii

9:30 A.M. Subject: *Urban Structure and Design*

Chairman: (to be announced)

Papers: 1. The City of the Future: The Problem of Optimal Design
Britton Harris, University of Pennsylvania

2. Intra-Urban Industrial Location Model—Design and Implementation
Stephen H. Putman, CONSAD Research Corporation

Discussants: *John F. Kain*, Harvard University; *Ira S. Lowry*, Massachusetts Institute of Technology

---

12:30 P.M. Lunch Session; Subject: *Reports on Regional Science Training and Research Programs*

Chairman: *Benjamin H. Stevens*, University of Pennsylvania

Reports: 1. U.C.L.A.

2. Syracuse University

3. Washington University

4. University of Pennsylvania

5. Harvard University

2:30 P.M. Subject: *Models for Regional Policy*

Chairman: *Julian Wolpert*, University of Pennsylvania

Papers: 1. Regional Allocation of Social Overhead Investment
Noboru Sakashita, Tohoku University

2. An Econometric Study of Regional Multipliers
Sung J. Kim, University of Pennsylvania

Discussion from the Floor

5:00 P.M.–6:00 P.M. Business Meeting

7:00 P.M. Dinner and Presidential Address

Chairman: *Charles L. Leven*, Washington University

Address: Global Science and the Tyranny of Space
William Warntz, American Geographical Society

SUNDAY, NOVEMBER 6, 1966

8:20 A.M.–9:20 A.M. Early-Bird Session: *Ph.D. Papers*

Chairman: *Waldo Tobler*, University of Michigan

Papers:
1. Transportation and the Growth of the São Paulo Economy
Howard L. Gauthier, Ohio State University
2. Travel in Urban Areas: Surveillance of a Special Panel and Longitudinal Analysis
Richard D. Worrall, Northwestern University
3. An Application of a Benefit-Cost Framework of Analysis to Selected Urban Redevelopment Projects in Indianapolis, Indiana: A Case Study of Locally Financed Redevelopment
Stephen D. Messner, University of Connecticut

9:30 A.M. Subject: *Resource Development in a Multi-Region Setting*

Chairman: *Lowden Wingo*, Resources for the Future

Papers:
1. Optimal Scheduling of Investments in a Multi-Regional Setting
Emilio Cassetti, University of Toronto
2. An Inter-Regional Demand-Supply Model for Water Resources Development
Nathaniel Wollman, University of New Mexico

Discussants: *David F. Bramhall*, Johns Hopkins University; Morgan Thomas, University of Washington

---

12:30 P.M. Lunch Session; Subject: *The EDA Summer Institute: An Evaluation*

Chairman: Anthony Pascal, U.S. Department of Commerce

2:30 P.M. A. *Section on Applied Regional Economics*

Chairman: *Benjamin Chinitz*, U.S. Department of Commerce

Papers:
1. Program Allocation in an Operational Setting
Robert Rauner and James Burns, Office of Program Evaluation, U.S. Department of Commerce
2. Regional Economic Development: Institution Building and Program Initiation
Edward F. R. Hearle, Office of Regional Economic Development, U.S. Department of Commerce
3. Planning for Growth in Appalachia
Monroe Newman, Pennsylvania State University

Discussants: *Charles M. Tiebout*, University of Washington; *Frederick Bell*, Boston Federal Reserve Bank; *Gordon Cameron*, University of Glasgow and University of Pittsburgh

---

2:30 P.M. B. Late-Bird Session: *Ph.D. Papers*

Chairman: *J. W. Milliman*, Indiana University

Papers:

1. Location and Its Relation to Output: A Sensitivity Analysis
   Gilbert A. Churchill, Jr., University of Wisconsin
2. Business Location Decisions by Entrepreneurs in Mexico City: Relationships between These Decisions and Perceptions of the Growth of the City by Decision Makers
   Robert D. Swartz, Northwestern University
3. Relationships between Driver Attitudes Toward Alternative Routes and the Characteristics of the Drivers and Routes
   Martin Wachs, Northwestern University
4. Simulation of the Urban Residential Site Selection Process: A Study of Taxonomic Problems and Decision Rules
   Joseph Stowers, Northwestern University
5. Critical Issues in Planning the Appalachian Development Highway System
   John M. Munro, Simon Fraser University

Fourteenth U.S. Annual Meetings
Cambridge, Massachusetts
November 3–5, 1967

FRIDAY, NOVEMBER 3, 1967

8:30 A.M. Registration

10:00 A.M. Words of Welcome, *Jose Luis Sert*, Dean, Graduate School of Design, Harvard University

---

10:15 A.M. Subject: *Sundry and Diverse Directions in Regional Science*

Chairman: *William A. Doeble*, Harvard University

Papers:

1. On the Linkage of Economic, Ecologic, and Other Natural Environment Systems for a Region
   Walter Isard, University of Pennsylvania

2. Spatial Theory and Human Behavior
Gunnar Ollson, University of Michigan

Discussant: *D. Michael Ray*, Spartan Air Services Limited

12:30 P.M. Lunch Session; Subject: *EDA in Action: Policies and Problems*

Chairman: *Anthony Pascal*, U.S. Department of Commerce

2:30 P.M. Subject: *Central Place Theory: Brand X and Brand Y*

Chairman: *William Alonso*, University of California (Berkeley)

Papers: 1. Interdependency of Spatial Structure and Spatial Behavior: A General Field Theory Formulation
Brian J. L. Berry, University of Chicago

2. Urban Structure and Regional Development
Gerald Hodge, University of Toronto

Discussant: *Gerald A. P. Carrothers*, Central Mortgage and Housing Corporation

---

5:00 P.M. Meeting of the Council of the Regional Science Association

---

8:00 P.M. Subject: *Multi-Regional Analysis*

Chairman: *John F. Kain*, Harvard University

Papers: 1. Optimal Investment in a Multi-Regional Setting via Recursive Programming
Emilio Casetti, Ohio State University

2. A Multi-Region, Multi-Sector Model of Equilibrium Growth
Koichi Mera, Harvard University

Discussant: *C. Peter Rydell*, Hunter College

8:00 P.M. Late-Bird Session: *Ph.D. Papers*
(concurrent session)

SATURDAY, NOVEMBER 4, 1967

7:30–9:20 A.M. Early-Bird Session: *Ph.D. Papers*

9:30 A.M. Subject: *Location & Geometry*

Chairman: *Leon Moses*, Northwestern University

Papers: 1. Location Theory Revisited
Benjamin H. Stevens, University of Pennsylvania

2. The Role of Topology and Geometry in Optimal Network Design
Christian Werner, Northwestern University

Discussant: *Waldo Tobler*, University of Michigan

12:30 P.M. Lunch Session; Subject: *Reports on Regional Science Training and Research Programs*

Chairman: *Benjamin H. Stevens*, University of Pennsylvania

Reports: Syracuse University
Washington University
University of Pennsylvania
Harvard University

2:30 P.M. Subject: *Intra-Urban Flows*

Chairman: *Lloyd Rodwin*, MIT

Papers: 1. The Urban Transportation Problem or, The Relationship of Transient Queing Behavior to Capacity Restraint Functions and to Travel Forecasting: An Introduction
Martin Wohl, RAND Corporation

2. Provision of Local Public Service
Michael B. Teitz, University of California (Berkeley)

Discussants: *Lee Cole*, Harvard University; *Julian Wolpert*, University of Pennsylvania

---

5:00–6:00 P.M. Business Meeting

7:00 P.M. Dinner and Presidential Address

Chairman: *William Warntz*, Harvard University

Address: *Britton Harris*, University of Pennsylvania

SUNDAY, NOVEMBER 5, 1967

7:30–9:20 A.M. Early-Bird Session: *Ph.D. Papers*

9:30 A.M. Subject: *Projection and Planning Models*

Chairman: *Robert W. Eisenmenger*, Federal Reserve Bank of Boston

Papers: 1. A Projection Framework for State Planning
David F. Bramhall, University of Pittsburgh

2. The Demographic Model: Empirical and Policy Aspects
James Beshers, MIT

Discussant: *John H. Cumberland*, University of Maryland

2:30 P.M. Subject: *Panel on Water Resources*

Chairman: *Maynard M. Hufschmidt*, University of North Carolina

Papers: 1. Analysis of the Integration of Regional Electric Power Systems in Columbia
Henry D. Jacoby, Harvard University

2. Planning Public Investment to Implement Regional Objectives: The Appalachian Water Resources Case
Charles A. Berry, U.S. Corps of Engineers and University of Cincinatti

Discussants: (to be announced)

2:30 P.M. Late-Bird Session: Ph.D. Papers
(concurrent session)

Appendix F

Regional Science Association Council Members

1957–58

President	Walter Isard			
President-elect	Walter Isard			
Vice President	Joseph L. Fisher			
" "	Rupert Vance			
Secretary	Robert A. Kavesh	(57–58)	(58–59)	(59–60)
Treasurer	Benjamin H. Stevens	"	"	"
Councillor	Edward A. Ackerman	(57–58)	(58–59)	(59–60)
"	Harvey S. Perloff	"	"	"
"	Seymour E. Harris	(57–58)	(58–59)	
"	Edward L. Ullman	"	"	
"	Louis B. Wetmore	(57–58)		
"	Stefan H. Robock	"		
Past President	- - - - - - - - - - - - -			

****

1958–59

President	Walter Isard			
President-elect	Robert B. Mitchell			
Vice President	Stefan H. Robock			
" "	Louis B. Wetmore			
Secretary	Robert A. Kavesh	(57–58)	(58–59)	(59–60)
Treasurer	Benjamin H. Stevens	"	"	"
Councillor	John A. Guthrie	(58–59)	(59–60)	(60–61)
"	Allan Rodgers	"	"	"

Councillor	Edward A. Ackerman	(57–58)	(58–59)	(59–60)
"	Harvey S. Perloff	"	"	"
"	Seymour E. Harris	(57–58)	(58–59)	
"	Edward L. Ullman	"	"	
Past President	Walter Isard			

****

1959–60

President	Robert B. Mitchell			
President-elect	Edward L. Ullman			
Vice President	J. Douglas Carroll, Jr.			
" "	Edgar M. Hoover			
Secretary	Robert A. Kavesh	(57–58)	(58–59)	(59–60)
Treasurer	Benjamin H. Stevens	"	"	"
Councillor	Coleman Woodbury	(59–60)	(60–61)	(61–62)
"	Hans Blumenfeld	"	"	"
"	John A. Guthrie	(58–59)	(59–60)	(60–61)
"	Allan Rodgers	"	"	"
"	Edward A. Ackerman	(57–58)	(58–59)	(59–60)
"	Harvey S. Perloff	"	"	"
Past President	Walter Isard			

****

1960–61

President	Edward L. Ullman			
President-elect	Edgar M. Hoover			
Vice President	F. S. Chapin, Jr.			
" "	William L. Garrison			
Secretary	Duane F. Marble	(60–61)	(61–62)	(62–63)
Treasurer	James Ganschow	"	"	"

Councillor	Joseph L. Fisher	(60–61)	(61–62)	(62–63)
"	John H. Nixon	"	"	"
"	Coleman Woodbury	(59–60)	(60–61)	(61–62)
"	Hans Blumenfeld	"	"	"
"	John A. Guthrie	(58–59)	(59–60)	(60–61)
"	Allan Rodgers	"	"	"
Honorary Chairman	Walter Isard			
Past President	Robert B. Mitchell			

****

1961–62

President	Edgar M. Hoover			
President-elect	William L. Garrison			
Vice President	Walter Christaller			
" "	Harold H. McCarty			
Secretary	Duane F. Marble	(60–61)	(61–62)	(62–63)
Treasurer	James Ganschow	"	"	"
Councillor	John Cumberland	(61–62)	(62–63)	(63–64)
"	Andrzej Wrobel	"	"	"
"	Joseph L. Fisher	(60–61)	(61–62)	(62–63)
"	John H. Nixon	"	"	"
"	Coleman Woodbury	(59–60)	(60–61)	(61–62)
"	Hans Blumenfeld	"	"	"
Honorary Chairman	Walter Isard			
Past President	Edward L. Ullman			

****

1962–63

President	William L. Garrison
President-elect	William L. C. Wheaton

Vice President	Torsten Hägerstrand			
" "	Leon N. Moses			
Secretary	Duane F. Marble	(60–61)	(61–62)	(62–63)
Treasurer	James Ganschow	"	"	"
Councillor	Charles Tiebout	(62–63)	(63–64)	(64–65)
"	Melvin M. Webber	"	"	"
"	John Cumberland	(61–62)	(62–63)	(63–64)
"	Andrzej Wrobel	"	"	"
"	Joseph L. Fisher	(60–61)	(61–62)	(62–63)
"	John H. Nixon	"	"	"
Honorary Chairman	Walter Isard			
Past President	Edgar M. Hoover			

****

1963–64

President	William L. C. Wheaton			
President-elect	Charles L. Leven			
Vice President	Chauncey Harris			
" "	Henry Fagin			
Secretary	Joseph Airov	(63–64)	(64–65)	(65–66)
Treasurer	David Bramhall	"	"	"
Councillor	Lowden Wingo, Jr.	(63–64)	(64–65)	(65–66)
"	Duane F. Marble	"	"	"
"	Charles Tiebout	(62–63)	(63–64)	(64–65)
"	Melvin M. Webber	"	"	"
"	John Cumberland	(61–62)	(62–63)	(63–64)
"	Andrzej Wrobel	"	"	"
Honorary Chairman	Walter Isard			
Past President	William L. Garrison			

****

1964–65

President	Charles L. Leven			
President-elect	William Warntz			
Vice President	John Dyckman			
" "	Kazimierz Dziewonski			
Secretary	Joseph Airov	(63–64)	(64–65)	(65–66)
Treasurer	David Bramhall	"	"	"
Councillor	Harold Barnett	(64–65)	(65–66)	(66–67)
"	Benjamin H. Stevens	"	"	"
"	Lowden Wingo, Jr.	(63–64)	(64–65)	(65–66)
"	Duane F. Marble	"	"	"
"	Charles Tiebout	(62–63)	(63–64)	(64–65)
"	Melvin M. Webber	"	"	"
Honorary Chairman	Walter Isard			
Past President	William L. C. Wheaton			

****

1965–66

President	William Warntz			
President-elect	Britton Harris			
Vice President	Benjamin Chinitz			
" "	Melvin Webber			
Secretary	Joseph Airov	(63–64)	(64–65)	(65–66)
Treasurer	David Bramhall	"	"	"
Councillor	William A. Alonso	(65–66)	(66–67)	(67–68)
"	Waldo R. Tobler	"	"	"
"	Benjamin H. Stevens	(64–65)	(65–66)	(66–67)
"	Harold Barnett	"	"	"
"	Lowden Wingo, Jr.	(63–64)	(64–65)	(65–66)

Councillor	Duane F. Marble	(63–64)	(64–65)	(65–66)
Honorary Chairman	Walter Isard			
Past President	Charles L. Leven			

****

1966–67

President	Britton Harris			
President-elect	Benjamin H. Stevens			
Vice President	Brian J. L. Berry			
" "	Gerald A. P. Carrothers			
Secretary	Sol Rabin			
Treasurer	Philip Hammer			
Councillor	Morgan Thomas	(66–67)	(67–68)	(68–69)
"	Andrei Rogers	(")	(")	(")
"	William A. Alonso	(65–66)	(66–67)	(67–68)
"	Waldo R. Tobler	(")	(")	(")
"	Harold Barnett	(64–65)	(65–66)	(66–67)
"	Ira S. Lowry	(")	(")	(")
Honorary Chairman	Walter Isard			
Past President	William Warntz			

****

1967–68

President	Benjamin H. Stevens	
President-elect	Torsten Hagerstrand	
Vice President	Edgar S. Dunn	
" "	John Friedmann	
Secretary	Sol Rabin	
Treasurer	Philip Hammer Jr.	
Councillor	Michael F. Dacey	(to 69–70)
"	Melvin L. Greenhut	(")

Councillor	Morgan Thomas	(to 68–69)
	Andrei Rodgers	(to 68–69)
"	William A. Alonso	(to 67–68)
"	Waldo R. Tobler	(")
Honorary Chairman	Walter Isard	
Past President	Britton Harris	

****

1968–69

President	Torsten Hagerstrand	
President-elect	Benjamin Chinitz	
Vice President	Roland Artle	
" "	Maynard Hufschmidt	
Secretary	Sol Rabin	
Treasurer	Philip Hammer Jr.	
Councillor	Genpachiro Konno	(to 70–71)
"	Thomas A. Reiner	(")
"	Michael F. Dacey	(to 69–70)
"	Melvin L. Greenhut	(")
"	Morgan Thomas	(to 68–69)
"	Andrei Rodgers	(")
Honorary Chairman	Walter Isard	
Past President	Benjamin H. Stevens	

References

Arrow, Kenneth J. and Gerard Debreu. 1954. Existence of an Equilibrium for a Competitive Economy, *Econometrica*, Vol. 22, No. 3, July, pp. 265-90.

Carroll, J. D. 1955. Spatial Interaction and the Urban-Metropolitan Description. *Papers and Proceedings of the Regional Science Association*, Vol. 1.

Carrothers, Gerald A. P. 1956. An Historical Review of the Gravity and Potential Concepts of Human Interaction. *Journal of American Institute of Planners*, Vol. 22, Winter.

Christaller, Walter. 1935. *Die Zentralen Orte in Süddeutschland.* Jena: G. Fischer.

Dunn, Edgar S. 1954. *The Location of Agricultural Production.* Gainesville: University of Florida Press.

Evans, W. Duane and Marvin Hoffenberg. 1952. *The Interindustry Relations Study for 1947.* Washington, D.C.: U.S. Department of Labor.

Hansen, Alvin, H. 1951. *Business Cycles and National Income.* New York: W. W. Norton.

Hansen, Niles. 1987. Poles of Development, *The New Palgrave Dictionary of Economics*. London: Macmillan.

Isard, Walter. 1942. Transport Development and Building Cycles. *Quarterly Journal of Economics*, November, pp. 90–112.

Isard, Walter. 1948. Some Location Factors in the Iron and Steel Industry Since the Early Nineteenth Century. *Journal of Political Economy*, Vol. LVI, June, pp. 203–17.

Isard, Walter. 1949. The General Theory of Location and Space Economy, *Quarterly Journal of Economics*, Vol. 63, no. 4, pp. 476–506.

Isard, Walter. 1956. *Location and Space Economy: A General Theory Relating to Industrial Location, Market Areas, Land Use, Trade and Urban Structure.* Cambridge, MA: MIT Press.

Isard, Walter. 1960. The Scope and Nature of Regional Science, *Papers and Proceedings of the Regional Science Association*, Vol. VI, pp. 9–34.

Isard, Walter. 1996. *Commonalities in Art, Science and Religion: An Evolutionary Approach.* Aldershot, England, Brookfield, Vt., USA: Avebury.

Isard, Walter and William M. Capron. 1949. The Future Locational Pattern of Iron and Steel Production in the United States, *Journal of Political Economy*, Vol. LVII, April, pp. 118–33.

Isard, Walter and Yun Ho Chung. 2001. Art and Science in Nuancing Conflict, with Particular Reference to Developed and Developing Nations, *Peace Economics, Peace Science and Public Policy*, Vol. 7, No. 3, Summer, pp. 29–35.

Isard, Walter and John N. Cumberland. 1950. New England as a Possible Location for an Integrated Iron and Steel Works. *Economic Geography*, Vol. 26, October, pp. 245–59.

Isard, Walter and Caroline Isard. 1943. The Transport-Building Cycle in Urban Development: Chicago. *The Review of Economic Statistics*, Vol. XXV, No. 4, November, pp. 224–226

Isard, Walter and D. Ostroff. 1960. General Interregional Equilibrium, *Journal of Regional Science*, Vol. 11, no. 1, pp. 67–74.

Isard, Walter and Merton J. Peck. 1954. Location Theory and Trade Theory: Short-Run Analysis, *Quarterly Journal of Economics*, Vol. LXVIII, May, pp. 305–320.

Isard, Walter et al. 1960. *Methods of Regional Analysis: An Introduction to Regional Science*. Cambridge, MA.: MIT Press.

Isard, Walter and John N. Cumberland (eds.). 1961. *Regional Economic Planning: Techniques of Analysis for Less Developed Areas*. Paris: Organization for European Economic Cooperation.

Isard, Walter et al. 1972. *Ecologic-Economic Analysis for Regional Development*. New York: Free Press.

Isard, Walter et al. 1998. *Methods of Interregional and Regional Analysis*. Aldershot, England, Brookfield, Vt., USA: Ashgate.

Keynes, John Maynard. 1936. *General Theory of Employment, Interest and Money*. New York: Harcourt, Brace and Co.

Lösch, August. 1940 and 1944. *Die räumliche Ordnung der Wirtschaft*. Jena: G. Fischer, 1st and 2nd editions.

Markusen, Ann. 2002. Two Frontiers for Regional Science. *Papers in Regional Science*, Vol. 81, No. 2.

Ohlin, Bertil. 1933. *Interregional and International Trade*. Cambridge, MA: Harvard University Press.

Perroux, Francoise. 1956. Les Mesures des progrès économiques et l'idée d'économic progressive, I.S.E.A., Paris.

Roberts, Clayton and David Roberts. 1985. *A History of England.* Englewood Cliffs, New Jersey: Prentice-Hall, Vol. 2.

Rostow, Walt W. 1960. *The Stages of Economic Growth.* London: Cambridge University Press.

Rostow, Walt W. (ed.). 1963. *The Economics of Take-Off into Sustained Growth.* New York: St. Martins Press.

Samuelson, Paul A. 1955. *Economics*. New York: McGraw Hill.

Spiethoff, Arthur. 1925. Krisen. *Handwörterbuch der Staatswissenschaften.*

Tinbergen, Jan. 1939. *Statistical Testing of Building Cycle Theories*. Geneva, Switzerland: League of Nations.

Vorhees, A. M. 1955. A General Theory of Traffic Movement. *Proceedings of the Institute of Traffic Engineers*.

Whitney, Vincent H. 1957. Potential Contribution of Regional Science to the Field of Sociology, *Papers and Proceedings of the Regional Science Association*, Vol. III, pp. 24–28.